First E

Viva V   ...ıınation in
Undergraduate Medicine;
1000 Questions

Dr S. Steele

Academic Medical Press

Published by Academic Medical Press, a division of Academic Medical Consulting. Nottingham, UK.

Academic Medical Consulting.
Ebury Road, Carrington, Nottingham NG5 1BB
Somniare audemus

First published 2011

Whilst the advice and information in this book are believed to be true and accurate at the date of going to press, neither the author nor the publisher can accept any legal responsibility or liability for any errors or omissions that may be made.

Any websites referred to in this publication are in the public domain and their addresses are provided by Academic Medical Consulting for information only. Academic Medical Consulting disclaims any responsibility for the content.

ISBN 978-0-9566443-3-6

Further copies can be obtained from: http://www.lulu.com
                                               http://www.amazon.com

Preface to the first edition

There has been a trend in modern undergraduate medical education to integrate the pre-clinical and clinical phases of the degree to yield a more complete understanding of the patient and patient management. Consequently, and understandably, this has created a demand for revision texts that are similarly integrative. It is hoped that this book will be a step towards satisfying such a demand.

The questions presented here have been generated over the last decade whilst teaching medical students from a wide range of backgrounds, with a wide range of abilities and a wide range of career interests. Furthermore, the questions have been chosen to ensure a balanced and representative view of a typical undergraduate medical course and are intended for revision and active reinforcement of core material.

Ideally, the material in this book should be used by students who have covered much of the core medical curriculum but require more material to challenge their knowledge and understanding. The questions can be used by students working alone or by students testing each other in pairs.

Finally, I would like to thank the medical students who reviewed the material in this book for their constructive criticism and their invaluable help in its final form and content. Their enthusiasm for this book was both encouraging and gratifying.

Oxford, October 2011   Dr S. Steele

Medical Student's Lament:

What do I need to know?
Why do I need to know it?
How do I learn it?
When is the exam?

Sinclair Steele, 2009

Contents

Session 1

1) Which virus causes shingles?

2) Are macrophages cells of acute or chronic inflammation?

3) Which pulse must you first feel for, before taking a manual blood pressure?

4) Above what respiratory rate should you call someone tachypnoeic?

5) A patient is said to be feverish. What is the minimum temperature that you would expect to find when you examined the patient?

6) Which of the following is not a derivative of the embryological foregut? Oesophagus, stomach, pancreas, liver, gallbladder, proximal 2/3 of the transverse colon, spleen.

7) What does *MCV* stand for?

8) What blood group antigens are on the surface of the red blood cells of an individual with blood group O neg.?

9) What does *FBC* stand for?

10) A patient has white plaques on their tongue. What is the most likely infective cause?

11) Assuming a normal fertilization, on what day does

implantation usually occur?

12) Which tissue in the human body has the greatest store of creatinine?

13) Is skeletal muscle derived from mesoderm, ectoderm or endoderm?

14) At which dermatome level does the umbilicus reside?

15) On a typical blood agar plate, what colour is beta haemolysis?

16) What is the name for the sounds that you are listening for when taking blood pressure?

17) An individual has the genotype 45XO – which disease/disorder are they likely to have?

18) If you reach out with your hand to ask for money, is your hand pronated or supinated?

19) Which has uracil as a base – DNA or RNA?

20) What is the name of the first formed diploid cell as a result of sexual reproduction?

21) Name the process whereby the equivalent RNA is made from its DNA template.

22) Describe Virchow's triad.

23) Name four examples of emboli.

24) Into what does the epiblast develop?

25) What is the general name for spherically shaped bacteria?

Session 1 Answers

1) Which virus causes shingles?

Herpes zoster.

2) Are macrophages cells of acute or chronic inflammation?

Chronic inflammation.

Remember that they are the dominant cells in an apical TB caseating granuloma that may be present for decades.

3) Which pulse must you *first* feel for, before taking a manual blood pressure?

Radial pulse.

Subsequently the brachial pulse is palpated prior to auscultation.

4) Above what respiratory rate should you call someone tachypnoeic?

Greater than 20 breaths per minute.

5) A patient is said to be feverish. What is the minimum temperature that you would expect to find when you examined the patient?

37.7 degrees centigrade or above.

6) Which of the following is not a derivative of the embryological foregut? Oesophagus, stomach, pancreas, liver, gallbladder, proximal 2/3 of the transverse colon, spleen.

Proximal 2/3 of the transverse colon.

7) What does *MCV* stand for?

Mean cell volume. (Mean corpuscular volume is also an acceptable answer).

8) What blood group antigens are on the surface of the red blood cells of an individual with blood group O neg.?

None (this is the universal donor).

9) What does *FBC* stand for?

Full blood count.

10) A patient has white plaques on their tongue. What is the

most likely infective cause?

Candida albicans.

11) Assuming a normal fertilization, on what day does implantation usually occur?

Day 6 or 7.

12) Which tissue in the human body has the greatest store of creatinine?

Skeletal muscle.

13) Is skeletal muscle derived from mesoderm, ectoderm or endoderm?

Mesoderm.

14) At which dermatome level does the umbilicus reside?

T10.

15) On a typical blood agar plate, what colour is beta haemolysis?

Transparent yellow.

16) What is the name for the sounds that you are listening for when taking blood pressure?

Korotkoff sounds.

17) An individual has the genotype 45XO – which disease/disorder are they likely to have?

Turner's syndrome.

18) If you reach out with your hand to ask for money, is your hand pronated or supinated?

Supinated.

19) Which has uracil as a base – DNA or RNA?

RNA.

20) What is the name of the first formed diploid cell as a result of sexual reproduction?

Zygote.

21) Name the process whereby the equivalent RNA is made

from its DNA template.

Transcription.

22) Describe Virchow's triad.

Vessel wall injury, abnormal blood flow and increased coagulability.

23) Name four examples of emboli.

Thromboembolic embolus, Fat embolus, Air embolus, Nitrogen embolus, Bone marrow embolus, Cholesterol and Mycotic.

Mycotic emboli are infective emboli that are often candidal or aspergillomatous in origin.

24) Into what does the epiblast develop?

Ectoderm and derived structures such as skin.

25) What is the general name for spherically shaped bacteria?

Cocci.

Session 2

1) What is the name for the condition during which the pH of human blood is below 7.35?

2) If a patient loses a significant blood volume due to trauma, with the exception of cross-matched blood, what other fluids could be used to replace the lost volume?

3) Give an example of co-dominance in genetic inheritance.

4) How are karyotypes prepared?

5) What are the functions of the greater omentum?

6) Is the BRCA-1 gene dominant or recessive?

7) What is metaphase?

8) Which is the commonest type of pathogenic mutation in humans?

9) What is epidemiology?

10) What factors are assumed in applying the Hardy-Weinberg equation/expression?

11) Considering genetics, what is anticipation?

12) What is prevalence?

13) What is the innervation of the internal oblique muscle?

14) What are the characteristics of autosomal dominant inheritance?

15) What does SIDS stand for?

16) Explain what is meant by the phrase "multi-factorial disorder."

17) Describe non-disjunction and give an example.

18) Which four principles can be used to assess the quality of health information data?

19) Explain the phrase "white coat hypertension."

20) Define incidence.

21) If you have an X-linked dominant trait as a male, are your male offspring likely to be normal or affected and why?

22) What are transcription and reverse transcription?

Session 2 Answers

1) What is the name for the condition during which the pH of human blood is below 7.35?

Acidosis.

2) If a patient loses a significant blood volume due to trauma, with the exception of cross-matched blood, what other types of fluid could be used to replace the lost volume?

Crystalloid or colloid.

3) Give an example of co-dominance in genetic inheritance.

ABO blood groups is the commonest example offered.

4) How are karyotypes prepared?

Essentially metaphase mitotic cells are arrested by treatment with colchicine, separated, fixed and stained appropriately with a stain such as Giemsa.

5) What are the functions of the greater omentum?

Abdominal policeman. Prevents visceral and parietal layers from fusing. Wraps itself around inflamed organs and and seals them from the rest of the peritoneal cavity.

6) Is the BRCA-1 gene dominant or recessive?

Recessive.

7) What is metaphase?

The phase in mitosis or meiosis when chromosomes line up along the equator of the cell and are held in place by the spindle. The chromosomes are decondensed.

8) Which is the commonest type of pathogenic mutation in humans?

Missense mutations (47%)

9) What is epidemiology?

The study of disease in populations.

10) What factors are assumed in applying the Hardy-Weinberg equation/expression?

Large population. No migration in or out of the population. No selection of alleles. Random mating. Constant mutation rate.

11) Considering genetics, what is anticipation?

A disease that occurs at an earlier age with each passing generation and may be more severe with each generation.

12) What is prevalence?

The number of people with a condition or disorder in a defined population at a defined time.

13) What is the innervation of the internal oblique muscle?

T7-12 and L1.

14) What are the characteristics of autosomal dominant inheritance?

Every generation affected; both sexes affected equally; affected child has (at least) one affected parent; unaffected individuals produce normal offspring; affected parent produces normal: affected children in the ratio of 1:1; homozygotes are rare.

15) What does SIDS stand for?

Sudden infant death syndrome

16) Explain what is meant by the phrase "multi-factorial disorder."

A disorder whose aetiology involves multiple factors including *environmental* and *genetic* factors.

17) Describe non-disjunction and give an example.

Failure of separation of chromosomes in mitosis or meiosis. Down's syndrome = trisomy 21; Edwards syndrome = trisomy 18.

18) Which four principles can be used to assess the quality of health information data?

CARTA: Completeness, Accuracy, Representativeness, Timeliness and Accessibility

19) Explain the phrase "white coat hypertension."

Stress induced rise in blood pressure at the prospect of a blood pressure measurement by a medical professional.

20) Define incidence.

Number of new cases of a disease or disorder in a defined population in a defined period of time.

21) If you have an X-linked dominant disease trait as a male, are your male offspring likely to be normal or affected and why?

Normal. As a male you will pass on a normal Y chromosome to all male offspring. Because you can only pass on one sex chromosome to your male offspring, the X-linked disease trait will not be passed on.

22) What are transcription and reverse transcription?

Transcription refers to the in vivo conversion of DNA into its equivalent RNA sequence, in the nucleus. Reverse transcription is the in vivo conversion of RNA into its equivalent DNA. A well known example is the viral RNA conversion into DNA as part of the HIV lifecycle.

Session 3

1) What is granulation tissue?

2) Describe the range of sizes that can be seen under a light microscope.

3) Name five processes that raise the intra abdominal pressure.

4) In biochemistry or pharmacology, what does the term affinity mean?

5) What is the major function of acetylcholinesterase in the human body?

6) What muscles make up the anterior wall of the inguinal canal?

7) What is the plural of thrombosis?

8) What is a fibroadenoma?

9) How may a tumour spread?

10) What colour is *gamma* haemolysis?

11) Considering skeletal muscle contraction, describe the cause of the absolute refractory period.

12) Parson's description of the patient's sick role includes four major items. Name them.

13) Which hepatitis virus is a DNA virus?

14) Which is more superficial - Scarpa's fascia or Camper's fascia?

15) Name the protein that is defective in osteogenesis imperfecta.

16) Are plasma cells more prominent in acute or chronic inflammation?

17) Considering treatment decision-making, what are the three types of doctor-patient relationships?

18) What is the term for a heart rate over 100 beats per minute?

19) What is drug efficacy?

20) What is the name for the voluntary movement of the shoulder forwards?

21) Define health.

22) Describe the Nernst equation.

23) What is the current life expectancy of (a) men and (b) women, in the UK? (2011)

24) Name three common epidemiological measures of fertility.

25) What is the abbreviation for the quaternary structure of haemoglobin?

26) What is a kinase?

27) What is the function of phospholipase C?

Session 3 Answers

1) What is granulation tissue?

Granulation tissue is richly vascular regenerative and reparative tissue that is usually found at the base of a wound. Its contents include growth factors, fibroblasts, extracellular matrix, blood vessels, neutrophils and macrophages.

2) Describe the range of sizes that can be seen under a light microscope.

Objects approximately **1μm – 1mm** in maximum extent.

3) Name five processes that raise the intra abdominal pressure.

Micturition, Defaecation, Coughing, Sneezing, Parturition, Vomiting, Laughing, Lifting weights.

4) In biochemistry or pharmacology, what does the term *affinity* mean?

Most commonly it refers to how well or tightly bound a ligand (agonist or antagonist) is to a receptor. It can be formally measured by either an association constant or a dissociation constant.

5) What is the major function of acetylcholinesterase in the human body?

To break down and inactivate the signalling molecule **acetylcholine**.

6) What muscles make up the anterior wall of the inguinal canal?

External oblique aponeurosis and internal oblique laterally.

7) What is the plural of thrombosis?

Thromboses.

8) What is a fibroadenoma?

A "breast mouse." This is a benign tumour of the breast consisting of stromal and epithelial elements that usually occurs in young women.

9) How may a tumour spread?

A tumour can spread:

Haematogenously – via arteries or veins.

Lymphatically – via lymphatic vessels and lymph nodes.

By direct invasion – local extension of tumour.

Transcoelomically – across a cavity – often the abdominal cavity.

By seeding – an iatrogenic process usually involving surgical intervention that results in the disturbance and translocation of tumour cells.

10) What colour is *gamma* haemolysis?

No haemolysis has occurred. The culture plates continue to be **red**.

11) Considering skeletal muscle contraction, describe the cause of the absolute refractory period.

The period when Na^{2+} channels are inactive immediately after opening. So they cannot initiate another action potential. This results in an absolute refractory period.

12) Parson's description of the patient's sick role includes four major items. Name them.

a) Patients want to get well as quickly as possible.

b) Patients are allowed and expected to shed some of their usual responsibilities.

c) Patients should seek professional medical advice if needed.

d) Patients are unable to get better on their own.

13) Which hepatitis virus is a DNA virus?

Hepatitis B.

14) Which is more superficial - Scarpa's fascia or Camper's fascia?

Camper's fascia.

15) Name the protein that is defective in osteogenesis imperfecta.

Type 1 Collagen. A glycine substitution impairs protein folding.

16) Are plasma cells more prominent in acute or chronic inflammation?

Chronic inflammation.

17) Considering treatment decision-making, what are the three types of doctor-patient relationships?

Shared, paternalistic and informed.

18) What is the term for a heart rate over 100 beats per minute?
Tachycardia.

19) What is drug efficacy?
How well a drug activates its target receptor to cause a downstream effect. Different drugs can produce varying sizes of responses even if they bind to the same proportion of the available receptors; such drugs have different efficacies.

20) What is the name for the voluntary movement of the shoulder forwards?
Protraction.

21) Define health.
WHO: "A stage of complete mental, social and physical well-being, not merely the absence of illness or infirmity".

Alternatively:
A) Absence of illness and
B) Maintaining functional ability and
C) Health = Equilibrium and
D) Health = Freedom

22) Describe the Nernst equation.
$E_{ion} = (RT/ZF)\ln([ion]_{out}/[ion]_{in})$

23) What is the current life expectancy of (a) men and (b) women, in the UK? (2011)
77 years for men and 82 years for women.

24) Name three common epidemiological measures of fertility.
Crude birth rate (live births/1000 population).

General fertility rate (live births/1000 females aged 15-44).

Total fertility rate (mean number of children a woman would have if she experienced the same age-specific fertility rate).

25) What is the abbreviation for the quaternary structure of haemoglobin?

$\alpha_2\beta_2$. This is a tetramer of two alpha subunits and two beta subunits.

26) What is a kinase?

An enzyme that transfers a phosphate moiety to a target molecule from ATP.

27) What is the function of **phospholipase C?**

Phospholipase is an enzyme that breaks down (hydrolyses) phosphatidylinositol 4,5-bisphosphate, PIP_2, to diacylglycerol, DAG, and inositol 1,4,5-triphosphate, IP_3. This process is significant because it produces two physiologically important secondary messengers that are intracellular signalling molecules.

Session 4

1) Name the layers of the stomach wall in sequence, starting at the lumen.

2) Are the following organs derived from the foregut, midgut or hindgut? Spleen, liver and pancreas.

3) According to Parson's what is expected from the doctor's professional role?

4) In 2010 what was the infant mortality in the UK?

5) Name three ways in which socioeconomic status is commonly measured.

6) According to the Registrar General's classification of social class, what types of occupation are in social class I?

7) The following moieties all have a similar type of function. What is it? Ca^{2+}, cAMP, cGMP and diacylglycerol.

8) Define an agonist.

9) What is a paracrine hormone?

10) What is an endocrine hormone?

11) Considering a typical human cell, what are the usual intracellular and extracellular sodium concentrations?

12) What is a phosphatase?

13) In skeletal muscle, what is the function of the sarcoplasmic reticulum?

14) What is PKC (biochemistry)?

15) Describe the mechanism of action of lignocaine/lidocaine.

16) Moving from the superior to inferior aorta, what is the sequence of occurrence of these arteries?
- Inferior mesenteric artery
- Coeliac trunk
- Superior mesenteric artery

Session 4 Answers

1) Name the layers of the stomach wall in sequence, starting at the lumen.

Mucosa

Submucosa

Muscularis propria (or Muscularis externa)

Serosa.

2) Are the following organs derived from the foregut, midgut or hindgut? Spleen, liver and pancreas.

Foregut.

3) According to Parson's what is expected from the doctor's professional role?

- Application of a high degree of skill.
- Action for the welfare of the patient.
- Must be objective.
- Must be guided by rules of professional practice.

4) In 2010 what was the infant mortality in the UK?

4.85 per 1000 live births.

5) Name three ways in which socioeconomic status is commonly measured.

By occupation.

By education.

By assets.

6) According to the Registrar General's classification of social class, what types of occupation are in social class I?

Professionals.

7) The following moieties all have a similar type of function. What is it? Ca^{2+}, cAMP, cGMP and diacylglycerol.

Secondary messengers.

Second messengers.

8) Define an agonist.

A ligand that binds to the active site of the receptor to trigger the receptor to cause downstream effects. *An antagonist may bind to the same site but does not trigger the receptor.*

9) What is a paracrine hormone?

A paracrine hormone is a hormone that is released and acts locally. *It is not released into the circulatory system.*

10) What is an endocrine hormone?

A hormone that is secreted into the circulatory system to act a distance from its site of origin.

11) Considering a typical human cell, what are the usual intracellular and extracellular sodium concentrations?

150 mM extracellular

10 mM intracellular

12) What is a phosphatase?

An enzyme that removes a phosphate group from a phosphorylated target molecule (usually a protein) by hydrolysis.

13) In skeletal muscle, what is the function of the sarcoplasmic reticulum?

Calcium ion storage and release.

14) What is PKC (biochemistry)?

Protein kinase C. A signalling enzyme that phosphorylates target proteins at serine or threonine amino acid residues. Hence it plays a role in some phosphorylation cascades.

15) Describe the mechanism of action of lignocaine/lidocaine.

It binds to and antagonizes the plasma membrane bound sodium channel in nerve cells. The blockage of this channel prevents the influx of sodium required to develop an action potential. Thus the nerve cell cannot transmit a pain signal.

16) Moving from the superior to inferior aorta, what is the sequence of occurrence of these arteries?

- Inferior mesenteric artery
- Coeliac trunk

* Superior mesenteric artery

1) Coeliac trunk 2) Superior mesenteric artery 3) Inferior mesenteric artery.

Session 5

1) There are three types of inequalities in health care. What are they?

2) What is the relationship between social class and smoking?

3) Define absolute poverty and relative poverty.

4) What is the commonest cause of cancer death in both men and women?

5) What are the two commonest causes of macrocytic normochromic anaemia?

6) What is the commonest cancer in females?

7) Are males or females more likely to commit suicide?

8) What is a totipotent stem cell?

9) Under normal circumstances what are the minimum and maximum volumes of the stomach?

10) Is the pylorus proximally or distally situated in the stomach?

11) Where do gastric slow waves originate?

12) Name two types of food that decrease the rate of gastric emptying.

13) Considering the gastrointestinal tract, what is MMC and what is its function?

14) What is in gastric juice?

15) What is cimetidine and where does it act?

16) Where in the GI tract does protein digestion begin?

17) How is the gastric mucosal barrier renewed?

18) When do we stop making foetal haemoglobin?

19) Where is intrinsic factor secreted?

20) Which disease occurs if there is dysfunctional or absent intrinsic factor?

21) What is the term for a respiratory rate of over 20 breaths per minute?

22) What is bronchial breathing a sign of?

23) What is the surface marking for the posterior lower border of the lungs?

24) What would you expect to feel on the left side on the mid-clavicular line at the fifth intercostal space?

Session 5 Answers

1) There are three types of explanations for inequalities in health care. What are they?

Behavioural/cultural.

Materialist/neo-materialist.

Psychosocial.

2) What is the relationship between social class and smoking?

Higher socioeconomic status individuals tend to smoke less.

3) Define absolute poverty and relative poverty.

Absolute poverty; individuals fall below the minimum standards of food, shelter and clothing necessary to sustain life.

Relative poverty; individuals fall below acceptable standards of living which prevents them from participating in community life.

4) What is the commonest cause of cancer death in both men and women?

Primary lung cancer.

5) What are the two commonest causes of macrocytic normochromic anaemia?

B_{12} deficiency and folate deficiency.

6) What is the commonest cancer in females?

Primary breast cancer.

7) Are males or females more likely to commit suicide?

Males.

8) What is a totipotent stem cell?

A stem cell that is capable of differentiating into any cell type.

9) Under normal circumstances what are the minimum and maximum volumes of the stomach?

50 and 1500mls respectively.

10) Is the pylorus proximally or distally situated in the stomach?

Distally.

11) Where do gastric slow waves originate?

In the pacemaker zone.

12) Name two types of food that decrease the rate of gastric emptying.

Fatty foods, acidic foods and hypertonic liquids.

13) Considering the gastrointestinal tract, what is MMC and what is its function?

MMC = Migrating Motor Complex. These are powerful contractions that sweep along the entire length of the stomach. MMC is associated with pyloric dilation. It is believed to have a housekeeping role of sweeping undigested food into the small intestine.

14) What is in gastric juice?

Gastric juice contains water, hydrochloric acid, pepsinogens/gastric lipase, mucus, hormones and intrinsic factor.

15) What is cimetidine and where does it act?

It is an antihistamine drug. It is a H_2 receptor antagonist acting on parietal cells in the stomach.

16) Where in the GI tract does protein digestion begin?

In the stomach. Secreted inactive pepsinogen is converted into the active pepsin in the acidic medium of the stomach.

17) How is the gastric mucosal barrier renewed?

- Irritation causes increased release of prostaglandins, mucus and bicarbonate.
- Stimuli that raise gastric acid secretion also increase mucus and bicarbonate secretion.
- There is a high rate of cell division in gastric epithelium.

18) When do we stop making foetal haemoglobin?

Most synthesis stops approximately three months after being born.

19) Where is intrinsic factor secreted?

Parietal cells of the stomach.

20) Which disease occurs if there is dysfunctional or absent intrinsic factor?

Pernicious anaemia.

21) What is the term for a respiratory rate of over 20 breaths per minute?

Tachypnoea.

22) Classically, which disease is bronchial breathing a sign of?

Pneumonia.

23) What is the surface marking for the posterior lower border of the lungs?

T10.

24) What would you expect to feel on the left side on the mid-clavicular line at the 5th intercostal space?

Apex beat of the heart.

Session 6

1) What is cholecystitis and what is the commonest mechanism by which it occurs?

2) Where in the body will you find Kupffer cells and what is their function?

3) Name two places where you might find stratified squamous epithelium.

4) What are the monomers that make up lactose?

5) Currently (2011) which country has the greatest percentage of obese individuals?

6) Where would you feel the apex beat?

7) What are astrocytes?

8) Where in the body would you expect to find Rokitansky-Aschoff sinuses?

9) Anatomically speaking, how many liver segments are there?

10) Where in the human body would you expect to find ciliated pseudostratified columnar epithelium?

11) Name two essential fatty acids that humans require.

12) How would you describe the physique of a cachectic patient?

13) Under what normal physiological conditions can S2 be split?

14) What is a pseudocyst?

15) What is the Medusa's head?

16) What is at the centre of a hepatic lobule?

17) Is coffee drinking associated with pancreatic cancer?

18) Which are the semilunar valves?

19) Which of the following groups is *not* considered to be at risk of malnutrition in the Western World?
- Homeless
- Teenagers
- Elderly
- Obese
- Pregnant women

20) How much of an "average 70kg man" is water?

21) Considering the somatic sensory component of the nervous system, list the general modalities.

22) What is the minimum percentage stenosis that must be present in the carotid arteries for a carotid bruit to be audible?

23) What is the BMI definition of obesity?

24) What type of disorder is Pompe's disease?

25) Where could you find urothelium in the normal human body?

26) What is the major site of drug and hormone detoxification in the human body?

27) What does the acronym ERCP stand for?

28) Describe a motor unit.

29) Which example of protein energy malnutrition has an enlarged abdomen, marasmus or kwashiorkor?

30) What is the meaning of the phrase *basal metabolic rate*?

31) Why are case-control studies particularly useful?

32) Generally speaking where are Brunner's glands found in the body?

33) What is the average life of a human red blood cell?

34) Define a fistula.

35) In 2009 what percentage of the UK population was of ethnic minority origin?

36) According to the 1999 Nuffield Trust study, what percentage of adults entering UK hospitals were malnourished?

37) Name five types of essential nutrients.

38) What are the classical causes of bias in case-control studies?

39) What is the major histological feature of coeliac disease?

40) Where in the body is bilirubin glucuronide made?

41) Precisely where in the body is glucagon secreted?

Session 6 Answers

1) What is cholecystitis and what is the commonest mechanism by which it occurs?

Cholecystitis is inflammation of the gallbladder. It is often caused by a biliary stone blocking the cystic duct.

2) Where in the body will you find Kupffer cells and what is their function?

In the liver. They are macrophages and so have a phagocytic function.

3) Name two places where you might find stratified squamous epithelium.

Upper and mid oesophagus.

Skin.

Distal cervical epithelium.

Vulva.

Vagina.

4) What are the monomers that make up lactose?

Glucose and galactose.

5) Currently (2011) which country has the greatest percentage of obese individuals?

UK.

6) Where would you feel the apex beat?

Left side, midclavicular line and fifth intercostal space.

7) What are astrocytes?

Neuroglial cells (non-electrically excitable cells) that provide an anatomical and functional support framework for nerve cells and fibres.

8) Where in the body would you expect to find Rokitansky-Aschoff sinuses?

Gallbladder.

9) Anatomically speaking, how many liver segments are there?

Eight.

10) Where in the human body would you expect to find ciliated pseudostratified columnar epithelium?

Respiratory tract.

11) Name two essential fatty acids that humans require.

α-linolenic acid and linoleic acid.

12) How would you describe the physique of a cachectic patient?

Wasted, skeletal or cadaverous.

13) Under what normal physiological conditions can S2 be split?

During inspiration.

14) What is a pseudocyst?

A **non-epithelial lined** fluid filled space in the pancreas that includes enzymes, inflammatory factors and cell debris. *True cysts have an epithelial lining. Pseudocysts are often found incidentally by radiological investigation.*

15) What is the Medusa's head?

Caput medusae are a sign of portal hypertension.

16) What is at the centre of a hepatic lobule?

Hepatic vein (central vein).

17) Is coffee drinking associated with pancreatic cancer?

Coffee drinking is associated with pancreatic cancer. It is believed to be a confounding factor with the true causative link being through smoking. *Smokers like to drink coffee......*

18) Which are the semilunar valves?

Aortic and pulmonary valves.

19) Which of the following groups is *not* considered to be at risk of malnutrition in the Western World?

- Homeless
- Teenagers
- Elderly
- Obese
- Pregnant women

The obese.

20) How much of an "average 70kg man" is water?

42 litres (60% of his mass).

21) Considering the somatic sensory component of the nervous system, list the general modalities.

Pain

Temperature

Touch

Pressure

Proprioception

22) What is the minimum percentage stenosis that must be present in the carotid arteries for a carotid bruit to be audible?

Approximately 70% stenosis.

23) What is the BMI definition of obesity?

A BMI greater than 30 is defined as obese.

24) What type of disorder is Pompe's disease?

Glycogen storage disease.

or an inborn error of metabolism,

or a congenital metabolic disease,

or an inherited metabolic disease.

25) Where could you find urothelium in the normal human body?

Urological tract i.e. prostatic urethra, bladder, ureters and kidney.

26) What is the major site of drug and hormone detoxification in the human body?

Liver.

27) What does the acronym ERCP stand for?

Endoscopic retrograde cholangiopancreatography.

28) Describe a motor unit.

A single alpha-motorneurone and all the muscle fibres supplied by it. Each muscle contains thousands of such motor units.

29) Which example of *protein energy malnutrition* has an enlarged abdomen, marasmus or kwashiorkor?

Kwashiokor.

30) What is the meaning of the phrase *basal metabolic rate*?

The energy required to perform fundamental metabolic processes such as breathing, transport and cellular turnover – **whilst at rest**.

31) Why are case-control studies particularly useful?

- They are good for investigating rare diseases.
- They are cheaper than cohort studies.
- They are faster than cohort studies.

32) Generally speaking where are Brunner's glands found in the body?

In the duodenum.

33) What is the average life of a human red blood cell?

120 days. *By three months there has been an approximately 75% turnover of glycosylated haemoglobin in erythrocytes, so glycosylated haemoglobin is used as an index of glycaemic control in diabetics for the previous three months.*

34) Define a fistula.

A fistula is an open pathological passageway between two epithelium lined organs or vessels that normally do not connect directly.

35) In 2009 what percentage of the UK population was of ethnic minority origin?

10% or 6 million.

36) According to the 1999 Nuffield Trust study, what percentage of adults entering UK hospitals were malnourished?

40%. Interestingly, 50% left hospital malnourished.

37) Name five types of essential nutrients.

Water, Vitamins, Minerals, Essential Fatty Acids, Essential Amino Acids, Calories.

38) What are the classical causes of bias in case-control studies?

Selection bias, information bias, confounding bias.

39) What is the major histological feature of coeliac disease?

Villous atrophy or subtotal villous atrophy.

40) Where in the body is bilirubin glucuronide made?

This conjugated bilirubin is made in the liver.

41) Precisely where in the body is glucagon secreted?

Alpha cells in the Islets of Langerhans in the pancreas.

Session 7

1) Name four radiological investigations for examining the pancreatobilary system.

2) Which two hormones are most important in control and coordination of the starved state?

3) What is the name of the cycle that removes lactate from skeletal muscle?

4) Name at least two individuals in a double blind clinical trial who should not know treatment allocation.

5) Name a contrast agent that can be used to enhance features visualized on an X-ray image.

6) What is the synonym for *stones in the gallbladder*?

7) Name four substances that can be measured as part of a liver function test.

8) Which three enzymatic pathways comprise biochemical respiration?

9) Which pathway carries out oxidative phosphorylation?

10) Name three enzymes produced by the pancreas.

11) What are the advantages of ERCP over the alternative radiological procedures?

12) Is hexokinase predominantly present in the liver or the brain?

13) Where does the citric acid cycle occur?

14) What is the difference between *intention to treat analysis* versus *as treated* analyses?

15) Which is the commonest pancreatic malignancy?

16) Name two possible complications of a liver biopsy.

17) For every glucose molecule that goes through glycolysis what is the maximum number of ATP molecules that can be created?

18) What is beta oxidation?

19) What colour is bone on an X-ray image?

20) Classically speaking which enzyme is most raised in the plasma in acute pancreatitis?

21) Which organ impairment has the following clinical features?
Acidosis
Muscle loss
Coagulopathy
Hepatorenal syndrome
Jaundice
Portal hypertension

22) What does V_{max} mean?

23) What is the name for the main enzyme that controls the rate of the Krebs Citric Cycle?

24) What is clinical equipoise?

25) What is the commonest initial site of primary pancreatic cancer?

26) What is NAFLD?

27) Which enzyme is the main control point for glycolysis?

28) Where does glycolysis occur?

29) Is the usual chest X-Ray *PA* or *AP*?

30) What colour is fat on an X-Ray image?

Session 7 Answers

1) Name four radiological investigations for examining the pancreatobilary system.

Ultrasound.

Endoscopic ultrasound.

CT.

ERCP, Endoscopic Retrograde Cholangiopancreatography.

MRCP, Magnetic Resonance Cholangiopancreatography.

PTC, Percutaneous Transhepatic Cholangiography.

2) Which two hormones are most important in control and coordination of the starved state?

Noradrenaline and cortisol.

3) What is the name of the cycle that removes lactate from skeletal muscle?

Cori cycle.

4) Name at least two individuals in a double blind clinical trial who should not know treatment allocation.

Any two of patient, clinician and assessor/analyzer.

5) Name a contrast agent that can be used to enhance features visualized on an X-ray image.

Barium sulphate.

6) What is the synonym for *stones in the gallbladder*?

Cholelithiasis. Cholecystitis is *inflammation* of the gallbladder.

7) Name four substances that can be measured as part of a liver function test.

Alanine transaminase, aspartate transaminase, gamma glutamyl transpeptidase, bilirubin, alkaline phosphatase and albumin.

8) Which three enzymatic pathways comprise biochemical respiration?

Glycolysis. Kreb's citric acid cycle. Electron transport chain/system.

9) Which pathway carries out oxidative phosphorylation?

Electron transport system/chain.

10) Name three enzymes produced by the pancreas.

The pancreas produces proteases, lipases and carbohydrate digesting enzymes. The latter can be enzymes that breakdown either polymeric sugars or disaccharides. Hence the list of pancreatic enzymes is extensive and includes:

Amylase, maltase, lactase, sucrase and phytase, trypsinogen/trypsin, chymotrypsinogen/chymotrypsin, carboxypeptidase, elastase, ribonuclease, lipase, cholesterol esterase and phospholipase A or B.

11) What are the advantages of ERCP over the alternative radiological procedures?

High sensitivity, high specificity and potentially therapeutic (dislodging stones).

12) Is hexokinase predominantly present in the liver or the brain?

Brain. *Glucokinase is present in the liver.*

13) Where does the citric acid cycle occur?

(Inner) mitochondrial matrix.

14) What is the difference between *intention to treat analysis* versus *as treated* analyses?

Intention to treat analysis includes all those in the trial whether or not they took the treatment (i.e. whether they were compliant). *As treated analyses* tend to yield a larger size of effect of the intervention – and so are more likely to yield statistically significant data.

Intention to treat analysis more realistically represents the behavior of a real population.

15) Which is the commonest pancreatic malignancy?

Primary pancreatic adenocarcinoma.

16) Name two possible complications of a liver biopsy.

Haemorrhage, pain, other organ perforation or death. (Death can be a complication of any major procedure or major disease).

17) For every glucose molecule that goes through glycolysis what is the maximum number of ATP molecules that can be created?

For every glucose molecule four ATP molecules are created. However two ATP molecules are consumed in glycolysis so there is a net yield of 2 ATP molecules.

18) What is beta oxidation?

The oxidation of fatty acids that occurs in the mitochondrion. This is part of respiration.

19) What colour is bone on an X-ray image?

White.

20) Classically speaking which enzyme is most raised in the plasma in acute pancreatitis?

Amylase.

21) Which organ impairment has the following clinical features?

Acidosis

Muscle loss

Coagulopathy

Hepatorenal syndrome

Jaundice

Portal hypertension

The liver.

22) What does V_{max} mean?

Maximum rate of enzymic catalysis.

23) What is the name for the main enzyme that controls the rate of the Krebs Citric Cycle?

Pyruvate dehydrogenase

24) What is clinical equipoise?

Clinical or real uncertainty regarding which of the compared treatments or managements is most effective.

25) What is the commonest initial site of primary pancreatic cancer?

Head of the pancreas.

26) What is NAFLD?

Non-alcoholic fatty liver disease.

A high fat cause diet can cause NAFLD.

27) Which enzyme is the main control point for glycolysis?

Phosphofructokinase.

28) Where does glycolysis occur?

In the cytoplasm (cytosol) of the cell.

29) Is the usual chest X-Ray *PA* or *AP*?

PA; partly to diminish size distortion of the heart.

30) What colour is fat on an X-Ray image?

Dark grey.

Session 8

1) What are the functions of the large intestine?

2) Where in the GI tract is most water reabsorbed?

3) What is the gastrocolic reflex?

4) Describe the Valsalva manoeuvre.

5) Describe the composition of faeces.

6) List the functions of bile salts.

7) What types of gallstones do you know?

8) Which pancreatic enzyme digests starch?

9) In what form is newly absorbed dietary fat carried in the lymphatics?

10) How is B_{12} carried in the duodenum, jejunum and ileum?

11) List three signs of malabsorption.

12) List the nine Bradford Hill's criteria that imply causality.

13) What is the name of the technique that uses bimanual palpation to feel a kidney?

14) What is the neurotransmitter at the skeletal muscle neuromuscular junction?

15) How many subunits are there in a normal G protein and which one binds GTP?

16) Name the four key phases of mitosis.

17) What is lactogen?

18) What is ANP and how does it work?

19) Name the amino acid that is used to biosynthesize dopamine, noradrenaline and adrenaline.

20) Name two hormones secreted by the posterior pituitary.

21) What is CRH?

22) What is LDL and what is its principle function?

23) One hormone can:
   a) Decrease muscle mass
   b) Increase bone resorption
   c) Inhibit inflammatory response and the immune response
Name the hormone.

24) Describe the key hormonal change in Cushing's syndrome.

25) Give two common causes of hyperthyroidism.

26) Name four physical features of Cushing's syndrome.

27) Where is the epigastric region?

28) What does the WHO regard as the key principles of palliative care?

Session 8 Answers

1) What are the functions of the large intestine?
   a) To extract sodium and water from the luminal contents.
   b) To make and store faeces.
   c) To move faeces towards the rectum.

2) Where in the GI tract is most water reabsorbed?

In the large intestine, specifically the colon.

3) What is the gastrocolic reflex?

The reflex occurs after the intake of a meal causes the stomach to be stretched. The stretch receptors are part of a reflex that increases the motility in the colon. This pushes the colonic contents into the rectum – triggering the urge to defaecate. The gastrocolic reflex is mediated by gastrin and extrinsic autonomic nerves.

4) Describe the Valsalva manoeuvre.

Full inspiration followed by forced expiration against a closed glottis. The diaphragm moves downwards, abdominal and thoracic muscles are contracted thus raising the intrabdominal pressure.

5) Describe the composition of faeces.

Faeces includes:
   • 75% **water**.
   • **Dead** and **living bacteria**.
   • Undigested material that includes **bile pigments** (stercobilin and urobilinogen) and **epithelial cells**.
   • **Fat**.
   • **Hydrogen disulphide, indole, skatole** and **thiols**.
   • **Inorganic matter**.

6) List the functions of bile salts.

Emulsification of dietary lipids.

Elimination of cholesterol.

Prevention of cholesterol precipitation in the gall bladder.

Facilitation of the absorption of fat soluble vitamins.

7) What types of gallstones do you know?

Cholesterol stones are the commonest.

Pigmented stones contain bilirubin and calcium.

Mixed stones.

8) Which pancreatic enzyme digests starch?

Pancreatic amylase.

9) In what form is newly absorbed dietary fat carried in the lymphatics?

As lipoproteins, specifically chylomicrons.

10) How is B$_{12}$ carried in the duodenum, jejunum and ileum?

Bound to intrinsic factor.

11) List three signs of malabsorption.

Weight loss

Abdominal distension/Increased flatulence

Diarrhoea

Steatorrhoea/Azotorrhea

(Pernicious) anaemia

12) List the nine Bradford Hill's criteria that imply causality.

Strength of association

Specificity of association

Consistency of association

Temporal sequence

Dose response

Reversibility

Coherence of theory

Biological plausibility

Analogy

13) What is the name of the technique that uses bimanual palpation to feel a kidney?

Balloting.

14) What is the neurotransmitter at the skeletal muscle neuromuscular junction?

Acetylcholine.

15) How many subunits are there in a normal G protein and which one binds GTP?

A normal G protein contains three subunits, alpha, beta and gamma. The alpha subunit binds GTP.

16) Name the four key phases of mitosis.

Prophase, Metaphase, Anaphase and Telophase.

17) What is lactogen?

A growth hormone released during pregnancy.

18) What is ANP and how does it work?

Atrial natriuretic peptide. It causes vasodilation and increases renal excretion of sodium ions to decrease systemic blood pressure.

19) Name the amino acid that is used to biosynthesize dopamine, noradrenaline and adrenaline.

Tyrosine.

20) Name two hormones secreted by the posterior pituitary.

ADH (also called vasopressin).

Oxytocin.

21) What is CRH?

Corticotrophin Releasing hormone.

22) What is LDL and what is its principle function?

LDL is low density lipoprotein. It transports cholesterol in the blood.

23) One hormone can:

    a)   Decrease muscle mass.

    b)   Increase bone resorption.

    c)   Inhibit inflammation and the immune response.

Name the hormone.

Cortisol (also called hydrocortisone).

24) Describe the key hormonal change in Cushing's syndrome.

The hormonal change is an over production of cortisol that is usually due to an adenoma in either the pituitary or adrenal gland.

25) Give two common causes of hyperthyroidism.

Any two of:

Grave's disease.

Toxic multinodular goiter.

Exogenous iodine administration.

Pituitary adenoma (TSH secreting).

26) Name four physical features of Cushing's syndrome.

Central weight gain.

Thin skin.

Easy bruising.

Abdominal striae.

Moon face.

Buffalo hump.

Muscle wasting.

Facial acne.

27) Where is the epigastric region?

Medial to the right hypochondrium.

Medial to the left hypochondrium.

Superior to the umbilical region.

28) What does the WHO regard as the key principles of palliative care?

Death is natural and inevitable.

Death should not be hidden away.

The quality of remaining life should be maximized.

The manner in which a person the lives, until they die, is individual.

The individual should die with dignity and have a good death.

Session 9

1) A 49 year old man has an expansile and pulsatile mass deep to his umbilicus. What is the most likely diagnosis?

2) Which type of urological symptom is hesitancy?

3) What is the accepted pathogenesis of type 1 diabetes mellitus?

4) Where in the human body would you expect to find the Glut-3 transporter?

5) Is glucagon an anabolic or catabolic hormone?

6) Is insulin resistance a common feature of type 1 diabetes mellitus or type 2 diabetes mellitus?

7) Which has the stronger association with family history, type 1 or type 2 diabetes mellitus?

8) What are the physiological actions of angiotensin II?

9) Which part of the adrenal cortex produces aldosterone?

10) What effect does an Addisonian crisis have on systemic blood pressure?

11) What structures are visualized by karyotyping?

12) Name the three abdominal regions that are adjacent to the left hypochondrium.

13) On auscultation of a patient's heart an ejection systolic murmur is heard. What is the significance of this murmur?

14) Is primary adenocarcinoma of the lung more common at the periphery or at the perihilar region?

15) Of what condition is visible peristalsis a sign?

16) Name two causes of bloody diarrhoea.

Session 9 Answers

1) A 49 year old man has an expansile and pulsatile mass deep to his umbilicus. What is the most likely diagnosis?

Abdominal aortic aneurysm.

2) Which type of urological symptom is hesitancy?

An obstructive symptom.

3) What is the accepted pathogenesis of type 1 diabetes mellitus?

An autoimmune attack on beta cells of the Islets of Langerhans in the pancreas.

4) Where in the human body would you expect to find the Glut-3 transporter?

The brain (in the cell membranes).

5) Is glucagon an anabolic or catabolic hormone?

Catabolic.

6) Is insulin resistance a common feature of type 1 diabetes mellitus or type 2 diabetes mellitus?

Type 2. Insulin resistance is a precursor to impaired glucose tolerance ("glucose intolerance").

7) Which has the stronger association with family history, type 1 or type 2 diabetes mellitus?

Type 2.

8) What are the physiological actions of angiotensin II?

Peripheral vasoconstriction.

Stimulation of vasopressin release due to an action at the hypothalamus.

Stimulation of secretion of aldosterone from the adrenal cortex; the aldosterone causes sodium reabsorption in the kidney and consequently reduces urine output.

9) Which part of the adrenal cortex produces aldosterone?

Zona glomerulosa.

10) What effect does an Addisonian crisis have on systemic blood pressure?

It causes a catastrophic drop in blood pressure.

11) What structures are visualized by karyotyping?

Chromosomes.

12) Name the three abdominal regions that are adjacent to the left hypochondrium.

Epigastric region.

Umbilical region.

Left lumbar region.

13) On auscultation of a patient's heart an ejection systolic murmur is heard. What is the significance of this murmur?

It is the murmur of aortic stenosis. Such murmurs increase in prevalence with age because of calcification of the aortic valve. This valvular disorder is a recognized cause of sudden death amongst adults.

14) Is primary adenocarcinoma of the lung more common at the periphery or at the perihilar region?

Periphery of the lung. More lung tissue is present at the periphery than at the hilum and as lung tissue is predominantly glandular, it is more likely that a glandular malignancy will occur at the periphery. Squamous cell carcinoma of the lung is more likely to occur at the hilar region because it classically develops through squamous metaplasia of respiratory epithelium (e.g. in the bronchus) caused by cigarette smoke.

15) Of what condition is visible peristalsis a sign?

Bowel obstruction.

16) Name two causes of bloody diarrhoea.

The three commonest causes are gastroenteritis, ulcerative colitis and Crohn's disease.

Session 10

1) Considering *Needs Assessment*, state the different types of needs of which you are aware.

2) Name three complications of diabetes mellitus.

3) Where would you expect to find the Glut-4 transporters?

4) Name three irritative urinary symptoms.

5) How does SOCRATES relate to your patient's description of pain?

6) Under normal circumstances what is the plasma osmolality?

7) What is macrosomia?

8) What is melaena and what is its significance?

9) Which animal insulin is most like human insulin?

10) Describe odynophagia.

11) What are the divisions of the autonomic nervous system?

12) What is the honeymoon period in endocrinology?

13) What is the classic site of radiation of pancreatic pain?

14) How does a sulphonylurea work?

15) Define symptom.

16) What is tenesmus?

17) Where would you expect to find GLUT-3 transporters?

18) Which artery supplies the majority of the hindgut?

19) Define homeostasis.

20) List six causes of abdominal distention.

21) Which cells in the body produce glucagon?

22) What is the typical cause of the radiological appearance called the "apple-core" deformity?

23) What does the acronym MODY stand for?

24) What is the nerve root of the umbilicus?

Session 10 Answers

1) Considering *Needs Assessment*, state the different types of needs of which you are aware.

Normative need – defined by experts.

Felt need – individual perception.

Expressed need – help seeking.

Comparative need – similar to those receiving help.

2) Name three complications of diabetes mellitus.

Ischaemic Heart Disease / Myocardial Infarction.

Stroke or Transient Ischaemic Attack.

Visual Impairment – diabetic retinopathy, glaucoma, cataracts.

Kidney Failure – diabetic nephropathy and nephrotic syndrome.

Nerve Damage – peripheral neuropathy, autonomic neuropathy.

Peripheral Vascular Disease – leg ulcers, gangrene.

Coma – diabetic ketoacidosis, hyperosmotic non-ketotic.

Infections.

Death.

There are also a large range of gestational complications including macrosomia and hyperglycaemia/gestational diabetes mellitus.

3) Where would you expect to find the Glut-4 transporters?

At skeletal muscle or adipose tissue.

4) Name three irritative urinary symptoms.

Any of:

Frequency

Urgency

Nocturia

Dysuria

5) How does SOCRATES relate to your patient's description of pain?

Site

Onset

Character

Radiation

Associated symptoms

Timing

Exacerbating or relieving factors

Severity

6) Under normal circumstances what is the plasma osmolality?

275 – 299 milliosmoles per kilogram

7) What is macrosomia?

Large foetus/neonate for gestational age.

8) What is melaena and what is its significance?

Black "tarry stools" that represent altered blood originating from an upper gastrointestinal haemorrhage.

9) Which animal insulin is most like human insulin?

Porcine insulin has greatest homology with human insulin.

10) Describe odynophagia.

Painful swallowing.

11) What are the divisions of the autonomic nervous system?

Sympathetic and parasympathetic.

12) What is the honeymoon period in endocrinology?

The diabetic is only partially "insulin dependent" because they still have sufficient functioning β-cells in the Islet of Langerhans to generate insulin if appropriately stimulated. The honeymoon period is over when the individual's synthesis and secretion of insulin becomes negligible.

13) What is the classic site of radiation of pancreatic pain?

Pancreatic pain often radiates to the back.

14) How does a sulphonylurea work?

It's action is to increase the secretion of insulin. It inhibits the K^+ channel in the β-cells of the Islets of Langerhans. This causes depolarisation of the cell membrane and opening of voltage dependent calcium channels. The calcium facilitates exocytosis and secretion of insulin.

15) Define symptom.

Any morbid phenomenon or *departure from normal* function, appearance or sensation, that is *experienced by the patient* and is *indicative of disease*.

16) What is tenesmus?

Feeling of incomplete evacuation after the bowels have been opened.

17) Where would you expect to find GLUT-3 transporters?

In the brain.

18) Which artery supplies the majority of the hindgut?

Inferior mesenteric artery.

19) Define homeostasis.

Homeostasis is the process by which an internal quantity is kept in physiological range through a dynamic equilibrium. The process requires sensors and feedback control.

20) List six causes of abdominal distention.

Fat

Faeces

Flatus

Fluid

Foetus

Flipping **tumour (or "Fatal-oma") i.e. a neoplasm**

21) Which cells in the body produce glucagon?

α-cells of the Islets of Langerhans in the pancreas.

22) What is the typical cause of the radiological appearance called the "apple-core" deformity?
Obstructing malignancy in the gastrointestinal tract (usually a colorectal adenocarcinoma).
23) What does the acronym MODY stand for?
Maturity onset diabetes of the young. This is similar to type 2 diabetes that occurs in older adults but MODY usually manifests in individuals under the age of 25 and has a strong hereditary contribution that shows autosomal dominant inheritance.
24) What is the nerve root of the umbilicus?
T10.

Session 11

1) Which nerve roots supply the ankle jerk reflex?

2) What are the ten components of the abbreviated mental test exam?

3) Precisely where in the body is T4 produced?

4) Name three causes of hypothyroidism.

5) Define *autocrine gland*.

6) The chromatids are lined up along the equator of the cell – what phase is this?

7) What is the best eye response on the Glasgow Coma Scale?

8) Which nerve roots supply the quadriceps or knee reflex?

9) What is the synonymous phrase for *mitosis promoting factor*?

10) What are the relevant equations/expressions that describe the Hardy-Weinberg equilibrium?

11) Name eight signs or symptoms of Cushing's disease.

12) Which has the greater calcium concentration, the extracellular fluid or the cytosol?

13) What is the MRC system?

14) Which glucose transporter would you expect to find in the erythrocyte cell membrane?

15) During a peripheral neurological examination which sensations should you test for?

16) During which phase in the cell cycle does DNA synthesis occur?

17) What is the genetic abnormality of Edward's syndrome?

18) What is the generic name for a hormone released by the pituitary gland that acts on a target gland to cause the hormonal effect?

19) In the mitochondrial inner membrane, electrons are passed from one membrane protein to another until they are finally accepted by oxygen. What is this process called?

20) In general terms are glucagon's actions catabolic or anabolic?

21) What is the best motor response on the Glasgow Coma Scale?

22) Considering metabolic endocrinology where would you expect to find voltage dependent calcium channels?

Session 11 Answers

1) Which nerve roots supply the ankle jerk reflex?

S1/S2

2) What are the ten components of the abbreviated mental test exam?

Asking the patient for their **age**.

Asking the patient for the **time**.

Testing the patient for the **recall of an address** just given to them.

Asking the patient for the **year**.

Asking the patient for the name of their current **location**. *Hospital or house number.*

Seeing if the patient can **recognise two individuals**.

Asking the patient for their **date of birth**.

Asking for the year of commencement of **World War I**.

Naming the present **monarch**/president/prime minister.

Counting backwards from 20.

3) Precisely where in the body is T4 produced?

Thyroid gland.

4) Name three causes of hypothyroidism.

Any of:

Overtreatment with anti-hyperthyroid drugs or radioisotopes such as iodine-131.

Iodine deficiency in the diet (can lead to goitre).

Surgical excision of the thyroid.

Malignancy in (or affecting) the thyroid.

Hashimoto's thyroiditis or other inflammatory/autoimmune diseases.

Pituitary lesion decreasing TSH secretion; commonly an adenoma.

Hypothalamic lesion decreasing TRH secretion; usually an adenoma.

Developmental abnormalities such as thyroid **aplasia** or hypoplasia.

Drugs e.g. **lithium.**

Biosynthetic metabolic disorders.

5) Define autocrine gland.

A gland that secretes a hormone which acts upon itself, the secreting gland.

6) The chromatids are lined up along the equator of the cell – what phase is this?

Metaphase.

7) What is the best eye response on the Glasgow Coma Scale?

Eye opens spontaneously.

8) Which nerve roots supply the quadriceps or knee reflex?

L3/4.

9) What is the synonymous phrase for *mitosis promoting factor*?

Maturation promoting factor.

10) What are the relevant equations/expressions that describe the Hardy-Weinberg equilibrium?

$p + q = 1$

$p^2 + q^2 + 2pq = 1$

11) Name eight signs or symptoms of Cushing's disease.

The complete list is extensive but the commonest answers are listed:

Truncal obesity/central weight gain.

Buffalo hump.

Moon face.

Depression.

Psychosis.

Insomnia.

Amenorrhea.

Poor libido.

Thin skin/poor wound healing.

Easy bruising.

Growth arrest.

Relative immunosuppression and infections.

Back pain/spinal deformities.

Hyperglycaemia/polyuria/polydipsia.

Abdominal striae.

Skin pigmentation.

Muscle wasting.

Hirsutism in women.

Acne.

Hypertension.

12) Which has the greater calcium concentration, the extracellular fluid or the cytosol?

Extracellular fluid.

13) What is the MRC system?

A clinical system for assessing the power of a patient's musculature.

The range is 0 – 5.

0 implies total paralysis.

5 implies normal power.

14) Which glucose transporter would you expect to find in the erythrocyte cell membrane?

GLUT-1

15) During a peripheral neurological examination which sensations should you test for?

All of: Light touch and fine touch

                   Pain

                   Vibration

                   Proprioception

                   Temperature

                   Pressure

16) During which phase in the cell cycle does DNA synthesis occur?

S phase.

17) What is the genetic abnormality of Edward's syndrome?
Trisomy 18.
18) What is the generic name for a hormone released by the pituitary gland that acts on a target gland to cause the hormonal effect?
Trophic hormones.
19) In the mitochondrial inner membrane, electrons are passed from one membrane protein to another until they are finally accepted by oxygen. What is this process called?
Electron transport system or electron transport chain.
20) In general terms are glucagon's actions catabolic or anabolic?
Catabolic.
21) What is the best motor response on the Glasgow Coma Scale?
Obeys commands.
22) Considering metabolic endocrinology where would you expect to find voltage dependent calcium channels?
In the cell membranes of the β-cells of the Islets of Langerhans in the pancreas. They help to control insulin secretion.

Session 12

1) Give two synonyms for the symptom of dyspepsia.

2) What is Trendelenburg's sign and what is its significance?

3) For which malignancy might you use the drug Herceptin?

4) Name the disease that is most likely to predispose to xanthelasma formation.

5) Where in the body would you expect to find the appendices epiploicae?

6) Which lymphoma is the Epstein Barr virus notorious for causing?

7) What is cytokinesis?

8) Which nerve roots supply the triceps reflex?

9) Where is McBurney's point?

10) If you wish to determine true leg limb length, between which points should it be measured?

11) Name five groups of head and/or neck lymph nodes.

12) What is the function of the SMAD pathway?

13) What is the function of the puborectalis muscle?

14) To which malignancy does human papillomavirus predispose women?

15) At what vertebral level is the xiphoid process?

16) Where in the human body would you expect to find taeniae coli?

17) What is an oncogene?

18) What kind of drug is Gleevec?

19) If on perusing a patient's notes you were to see three plus signs (+++) next to a reflex, what would it signify?

20) Why can alcohol abuse cause gynaecomastia?

21) Which nerve roots supply the parasympathetic innervation of the anal canal?

22) Name three tumour suppressor genes.

23) If you wished to confirm adequate peripheral blood flow to the feet, which arteries would you palpate?

24) If no mutation is present, does phosphorylation of retinoblastoma protein transmit or inhibit the growth signal?

Session 12 Answers

1) Give two synonyms for the symptom of dyspepsia.

A lay person (non-doctor) might describe the symptom of dyspepsia as **heartburn** or **indigestion**.

2) What is Trendelenburg's sign and what is its significance?

It is used to demonstrate weak hip abductors. The patient is asked to stand up and raise one foot above the floor for thirty seconds. Then repeat for the other leg. The side with weak abductors will drop the contralateral (opposing) leg.

3) For which malignancy might you use the drug Herceptin?

Primary breast carcinoma.

4) Name the disease that is most likely to predispose to xanthelasma formation.

Hypercholesterolaemia.

5) Where in the body would you expect to find the appendices epiploicae?

On the large intestine.

6) Which lymphoma is the Epstein Barr virus notorious for causing?

Burkitt's lymphoma.

7) What is cytokinesis?

Division of the cell cytoplasm.

8) Which nerve roots supply the triceps reflex?

C7/8.

9) Where is McBurney's point?

It is the point a third of the way along the straight line from the anterior superior iliac spine to the umbilicus. It is usually the site of origin of appendiceal pain in acute appendicitis.

10) If you wish to determine true leg limb length, between which points should it be measured?

Anterior superior iliac spine to medial malleolus.

11) Name five groups of head and/or neck lymph nodes.

Any five of: Supraclavicular

Submental

Submandibular

Pre-auricular

Post-auricular

Cervical

Anterior Triangle and *Posterior Triangle* are also acceptable answers.

12) What is the function of the SMAD pathway?

It is the signal transduction pathway (signalling pathway) for inhibitory growth factors i.e. it transmits the signal to prevent growth.

13) What is the function of the puborectalis muscle?

It forms the sling that results in the anorectal flexure. It helps to control exit of faeces.

14) To which malignancy does human papillomavirus predispose women?

Primary cervical carcinoma.

15) At which vertebral level is the xiphoid process?

T9

16) Where in the human body would you expect to find taeniae coli?

In the large intestine.

17) What is an oncogene?

An activated proto-oncogene whose presence can predispose to malignancy, usually by eliciting a continuous and unregulated growth signal. Oncogenes are *overexpressed* in the tissues that they are affecting to cause cancer.

18) What kind of drug is Gleevec?

A tyrosine kinase inhibitor drug. *Or*

A signal transduction inhibitor drug.

It has been used as a magic bullet to kill cancer cells.

19) If on perusing a patient's notes you were to see three plus signs (+++) next to a reflex, what would it signify?

An exaggerated reflex.

20) Why can alcohol abuse cause gynaecomastia?

The liver impairment that accompanies chronic alcohol abuse means that the liver is not as able to carry out its function of detoxifying and inactivating sex hormones.

21) Which nerve roots supply the parasympathetic innervation of the anal canal?

S2,3,4   Pelvic splanchnic nerve.

22) Name three tumour suppressor genes.

Common answers to this question include:

Retinoblastoma protein (Rb protein)

p53

p16

BRCA1

Wnt-APC

SMAD

BAX

23) If you wished to confirm adequate peripheral blood flow to the feet, which arteries would you palpate?

Dorsalis pedis artery.

Posterior tibial artery.

24) If no mutation is present, does phosphorylation of retinoblastoma protein transmit or inhibit the growth signal?

Transmit. The phosphorylated form releases E2F which acts at a transcriptional level to allow synthesis of mitotic proteins.

25) What is MPF?

Maturation promoting factor or mitosis promoting factor.

It is composed of cyclin B and Cdc2 (Cdc2 = CDK1).

Session 13

1) Starting from the luminal aspect, what are the three histological layers of either arteries or veins?

2) What is orthopnea?

3) What is another name for the paramesonephric tissue that goes on to form female genitalia?

4) A patient you know collapses in front of you. You suspect anaphylaxis. Assuming you have access to the appropriate medications, what should you do?

5) Where in the body is the olecranon fossa?

6) What is health psychology?

7) What is the other name for the mesonephric tissue that will go on to form the male genitalia?

8) What is the difference between atheroma and atherosclerosis?

9) Name three different types of necrosis.

10) Which has the higher mean blood pressure, the lumen of the aorta or the lumen of the vena cava?

11) In which organ are you most likely to find seminiferous tubules?

12) Where in a bone would you find a diaphysis?

13) What is paroxysmal nocturnal dyspnea?

14) In immunology, what does HLA stand for?

15) What is the pathological difference (i.e. mechanism of disease) between myocardial infarction and angina?

16) Give two examples of type 1 hypersensitivity reactions.

17) List the key genotypic and phenotypic features of Turner's syndrome.

18) Describe two general mechanisms by which psychological processes can influence behaviour.

19) What is the surface anatomy marking for the posterior border of the lungs?

20) Which organ is particularly susceptible to damage by paracetamol?

21) What are the three components of the levator ani?

22) Describe Milgram's obedience study.

23) What is the plural of phalanx?

24) Name two anatomical differences between the right and the left lung.

25) Where does a Colles fracture occur?

26) What are the core components of the Calgary-Cambridge Guide for doctor-patient communication?

27) Which pelvic nerve may be damaged as a result of childbirth and what are its nerve roots?

28) What structures are found in a normal hepatic portal tract?

29) Name three points at which the oesophagus narrows.

30) Name four common types of cell injury.

31) Name three functions of the pelvic floor.

32) What does "fossa" mean?

33) What is mosaicism?

Session 13 Answers

1) Starting from the luminal aspect, what are the three histological layers of either arteries or veins?

Tunica intima, tunica media, tunica adventitia.

2) What is orthopnea?

Breathlessness at light when lying down; the patient has to prop themselves up with pillows. The severity of the orthopnea can be measured by the number of pillows required.

3) What is another name for the paramesonephric tissue that goes on to form female genitalia?

Mullerian duct or Mullerian tract or Mullerian tissue.

4) A patient you know collapses in front of you. You suspect anaphylaxis. Assuming you have access to the appropriate medications, what should you do?

Give 500µg adrenaline intramuscularly (to be repeated in 5 minutes if no response is elicited).

Give 10mg chlorpheniramine intravenously.

Give 10mg hydrocortisone intravenously.

Continue supportive management with the following, as appropriate:

Intravenous fluids.

High flow oxygen.

(Nebulised) salbutamol.

5) Where in the body is the olecranon fossa?

Distal posterior part of the humerus. (Or in the elbow).

6) What is health psychology?

The scientific study of psychological processes that influence health, illness and health care.

7) What is the other name for the mesonephric tissue that will go on to form the male genitalia?

Wolffian duct/tissue.

8) What is the difference between atheroma and atherosclerosis?

Atheroma = the deposit; Atherosclerosis = the disease process.

9) Name three different types of necrosis.

Any three of: Coagulative necrosis.

Liquefactive necrosis.

Fat necrosis.

Caseous necrosis.

10) Which has the higher mean blood pressure, the lumen of the aorta or the lumen of the vena cava?

Aorta.

11) In which organ are you most likely to find seminiferous tubules?

Testis.

12) Where in a bone would you find a diaphysis?

The centre (central part) of the shaft.

13) What is paroxysmal nocturnal dyspnea?

Acute respiratory distress that wakens a patient at night because of an underlying cardiac pathology (usually heart failure).

14) In immunology, what does HLA stand for?

Human Leucocyte Antigen.

15) What is the pathological difference (i.e. mechanism of disease) between myocardial infarction and angina?

MI = complete occlusion by blood clot (fast); Angina = partial occlusion by atheroma (slow).

16) Give two examples of type 1 hypersensitivity reactions.

Most students concentrate on *atopy*:

Anaphylaxis

Eczema/atopic dermatitis

Hayfever

Asthma

17) List the key genotypic and phenotypic features of Turner's syndrome.

Genotype XO (monosomy X).

Short stature.

Webbed neck.

Low-set ears.

No secondary female characteristics.

18) Describe two general mechanisms by which psychological processes can influence physical health.

Directly through changed psychological processes.

Indirectly through changed behavior.

19) What is the surface anatomy marking for the posterior lower border of the lungs?

T10, 10th rib.

20) Which organ is particularly susceptible to damage by paracetamol?

The liver.

21) What are the three components of the levator ani?

Puborectalis, pubococcygeus and iliococcygeus muscles.

22) Describe Milgram's obedience study.

Under instruction from an authority figure, the participants were asked to apparently give increasingly painful shocks to a "learner" whose screams could be heard. 65% of the participants delivered the highest shock level.

23) What is the plural of phalanx?

Phalanges.

24) Name two anatomical differences between the right and the left lung.

The right lung has three lobes whereas the left lobe has two lobes.

The right lung is larger. The left lung has the cardiac notch and the right bronchus has a straighter path that makes aspiration here more likely than in the left bronchus.

25) Where does a Colles fracture occur?

At the distal radius causing a dinner fork deformity, usually because of a fall onto an outstretched hand.

26) What are the core components of the Calgary-Cambridge Guide for doctor-patient communication?

Initiating the session.

Gathering information.

Providing structure.

Building the relationship.

Explanation and planning.

Closing the session.

27) Which pelvic nerve may be damaged as a result of childbirth and what are its nerve roots?

Pudendal nerve with nerve roots from S2, 3 and S4.

28) What structures are found in a normal hepatic portal tract?

Hepatic artery (hepatic arteriole).

Hepatic portal vein (hepatic venule).

Bile duct (bile ductule).

29) Name three points at which the oesophagus narrows.

Where the pharynx abuts the oesophagus.

Where the aortic arch abuts the oesophagus.

Where the oesophagus passes through the diaphragm.

30) Name four common types of cell injury.

Hypoxic, chemical (e.g. poisons), infective, heat/cold, immunological/inflammatory and nutritional insufficiency.

31) Name three functions of the pelvic floor.

Supports internal organs against gravity.

Allows increases in intra-abdominal pressure.

Allows passage and function of urethra, vagina and rectum.

32) What does "fossa" mean?

A shallow concavity in a bone. *Fossa is the latin word for ditch.*

33) What is mosaicism?

The situation where two populations of cells with different genotypes occur in one individual, that individual having developed from a single fertilized egg.

Session 14

1) What is an opsonin?

2) What are the two most important factors in determining the prognosis for any cancer (malignancy)?

3) What is cognitive psychology?

4) Name the muscles of the rotator cuff.

5) What does the S2 heart sound represent?

6) Which cell is most important in mediating acute inflammation?

7) What does "positive inotrope" mean?

8) What is the medical phrase synonymous with hayfever?

9) Name three possible causes of a transudate.

10) What type of epithelium lines the outermost part of the ectocervix?

11) What do the acronyms ESR and CRP stand for?

12) Name five adverse effects or complications of long term, high dose prednisolone use.

13) Which adrenergic receptor must be stimulated to counteract allergic bronchospasm?

14) What is the plural of foramen?

15) What is the name of the process whereby neutrophils migrate towards a chemical agent?

16) State another name for the rectouterine pouch.

17) Considering human psychology, classically what are the two major factors that determine behaviour?

18) What does the S1 heart sound represent?

19) Which metal is most likely to cause allergic contact dermatitis?

20) What effect does damage to the serratus anterior muscle have on the appearance of the scapula?

21) Where in the human body would you find the posterior fornix?

22) Considering BMI, what is the range of normality?

23) What is social psychology?

24) Which cells present CD4, cytotoxic T cells or T helper cells?

25) Which disease is associated with the antiphospholipid antibody?

26) Where in the body would you expect to find P-selectin?

27) What is the embryological origin of the heart?

28) Name the cranial nerves, in sequence. (No numerical naming).

29) Which arthritic disease classically shows an ulnar deviation deformity?

30) What type of joint is the acromioclavicular joint?

31) What type of valve disorder causes an ejection systolic murmur?

32) What is the minimum length of time that you must go without eating to achieve the starved state?

33) Name four bodies (organs or vessels) present in the mediastinum.

34) Which type of antibody is overproduced in atopic individuals?

35) Name the four major cardiac valves.

36) What single word means "enlarged heart"?

37) What five belief dimensions can shape a patient's illness representation?

Session 14 Answers

1) What is an opsinin?

A protein that coats and labels a substance for phagocytosis. Usually the substance is a foreign material such as a bacterium.

2) What are the two most important factors in determining the prognosis for any cancer (malignancy)?

Stage and grade.

3) What is cognitive psychology?

The scientific study of basic mental abilities. It is concerned with the acquisition, processing and storage of information. Accordingly it deals with memory, language processing, perception, problem solving and thinking.

4) Name the muscles of the rotator cuff.

SITS: Supraspinatus, infraspinatus, teres minor and subscapularis.

5) What does the S2 heart sound represent?

Aortic and pulmonary valve closure.

6) Which cell is most important in mediating acute inflammation?

Neutrophil.

7) What does "positive inotrope" mean?

A drug or hormone that increases the force of cardiac muscle contraction.

8) What is the medical phrase synonymous with hayfever?

Seasonal allergic rhinitis.

9) Name three possible causes of a transudate.

A common transudate is *pulmonary oedema*; this can be caused by renal failure, hepatic failure and cardiac failure. Ascites may also be a transudate and is often caused by hepatic impairment and the consequent portal hypertension. Generally, any process that lowers the plasma osmotic concentration can cause a transudate.

10) What type of epithelium lines the outermost part of the ectocervix?

(Non-keratinizing) stratified squamous cell carcinoma.

11) What do the acronyms ESR and CRP stand for?

ESR = Erythrocyte sedimentation rate.

CRP = C reactive protein.

Both are inflammatory markers.

12) Name five adverse effects or complications of long term, high dose prednisolone use.

Any five from any of the lists below:

Fat deposition (moon face, buffalo hump, central obesity).

Skin changes (poor healing, easy bruising, abdominal striae, thin skin, hirsutism).

Catabolic effects (loss of muscle bulk, osteoporosis/fractures and kidney stones, diabetes mellitus as a result of increased gluconeogenesis).

Immune effects (relative immunosuppression and infections).

Reproductive (decreased fertility, decreased sex drive, menstrual irregularity).

CNS effects (depression, emotional lability and psychosis).

Systemic effects (hypertension).

13) Which adrenergic receptor must be stimulated to counteract allergic bronchospasm?

β_2 adrenergic receptors are triggered by the adrenaline given for anaphylaxis.

14) What is the plural of foramen?

Foramina.

15) What is the name of the process whereby neutrophils migrate towards a chemical agent?

(Positive) chemotaxis.

16) State another name for the rectouterine pouch.

Pouch of Douglas.

17) Considering human psychology, classically what are the two major factors that reinforce behavior?

Reward and punishment.

18) What does the S1 heart sound represent?

Closing of the atrioventricular valves; mitral and tricuspid valves.

19) Which metal is most likely to cause allergic contact dermatitis?

Nickel.

20) What effect does damage to the serratus anterior muscle have on the appearance of the scapula?

Causes winging of the scapula.

21) Where in the human body would you find the posterior fornix?

Deep to the posterior lip of the cervix and anterior to the pouch of Douglas (rectouterine pouch).

22) Considering BMI, what is the range of normality?

$18.5 - 24.9$ kg/m^2

23) What is social psychology?

The scientific study of how people's thoughts, feelings and actions are influenced by their social environment.

24) Which cells present CD4, cytotoxic T cells or T helper cells?

T helper cells.

25) Which disease is associated with the antiphospholipid antibody?

SLE, Systemic lupus erythematosus.

26) Where in the body would you expect to find P-selectin?

Endothelial cells.

27) What is the embryological origin of the heart?

Splanchopleuric mesoderm.

28) Name the cranial nerves, in sequence. (No numerical naming)

Olfactory, optic, oculomotor, trochlear, trigeminal, abducens, facial, vestibulocochlear, glossopharyngeal, vagus, accessory, hypoglossal.

29) Which arthritic disease classically shows an ulnar deviation deformity?

Rheumatoid arthritis.

30) What type of joint is the acromioclavicular joint?

(Synovial) plane joint.

31) What type of valve disorder causes an ejection systolic murmur?

Aortic stenosis (aortic valve stenosis).

32) What is the minimum length of time that you must go without eating to achieve the starved state?

12 hours.

33) Name four bodies (organs or vessels) present in the mediastinum.

Answers may include:

Heart, thymus, trachea, oesophagus, ascending aorta, lymph nodes, thoracic duct, internal mammary artery and internal mammary vein.

34) Which type of antibody is overproduced in atopic individuals?

IgE.

35) Name the four major cardiac valves.

Mitral, tricuspid, pulmonary and aortic valves.

36) What single word means "enlarged heart"?

Cardiomegaly.

37) What five belief dimensions can shape a patient's illness representation?

Identity: what is it?

Cause: what caused it?

Time: how long will it last?

Consequence: how will it impact my life?

Control-Cure: can it be treated, controlled or managed?

Session 15

1) What does the phrase "positive chronotrope" mean?

2) What does the acronym *CLL* stand for?

3) Who usually develops CLL – young, middle aged or older people?

4) Which cells express CD8, helper T cells or cytotoxic T cells?

5) Name an inflammatory mediator responsible for vasodilation.

6) On a plain chest X-ray image, what structures make up the right border of the cardiac shadow?

7) What histological stain would you use to definitively identify mycobacterium tuberculosis?

8) Name four examples of chronic inflammatory diseases.

9) What is the plural of diverticulum?

10) As a physician, what clinical features cause you to think of the diagnosis of an immunodeficiency disease?

11) What is the single word that means "enlarged liver"?

12) If blood collects in the pericardial space, what is it called, and what are the possible consequences?

13) Name two pyrogens responsible for initiating fever in the human body.

14) Which antibody is more useful in identifying an acute infection, IgM or IgG?

15) The median nerve supplies cutaneous sensation to which fingers?

16) Which is more worrying in the breast, a mobile palpable mass or a tethered palpable mass?

17) Name three functions of macrophages.

18) Name two systemic disorders that can occur as result of multiple myeloma.

19) Classically, with which groups of people do doctors communicate poorly?

20) Which nerve passes through the spiral groove of the humerus?

21) What effects does radial nerve damage have on the wrist?

22) Numerically, what is the minimum BMI that is classed as obese?

23) What does the acronym PTCA stand for?

24) Name four behaviours by doctors that are notorious for blocking patient disclosure.

25) Which disease produces the paraprotein called Bence-Jones protein?

26) If you were to give the following diseases one unifying classification, what would it be? Selective IgA deficiency, severe combined immunodeficiency, Di George syndrome and Bruton's X-linked agammaglobinaemia.

27) Name two excised vessels that are used for coronary artery bypass grafts, CABG.

28) Name the four lobes of the liver.

29) What is myoglobin?

30) "A lower limit of 12g/dL in women and 14g/dL in men." To what is the speaker referring?

31) Give three examples of the health effects of poor communication between doctors and patients.

32) Which spinal nerve root supplies all of the intrinsic hand muscles?

33) From which nerve roots does the brachial plexus originate?

34) From which white blood cell is the macrophage derived?

35) What is the key cellular abnormality of Bruton's agammaglobinaemia?

36) Name four factors that affect the patient's perception of consultation quality.

37) Does the sympathetic innervation of the heart decrease or increase the heart rate?

38) On a plain chest X-ray radiograph, which structures make up the right border of the cardiac image?

39) Is the Ghon focus in the apex of the lungs an example of acute or chronic inflammation?

40) After a shoulder dislocation, which nerve function should you test?

41) Which muscles mark the anterior border of the axilla?

42) Which white blood cells are involved in chronic inflammation?

43) Which of the following cells are not directly involved in chronic inflammation; macrophages, plasma cells, lymphocytes or erythrocytes?

44) What is the minimum length of time that must pass after a meal before the fasted state is reached?

45) Which chronic inflammatory cell has a "clock face" nucleus?

46) Name three disorders that an atopic individual would be more likely to experience.

47) Explain the difference between a lymphoma and a leukaemia.

Session 15 Answers

1) What does the phrase "positive chronotrope" mean?

A drug or hormone that increases the rate of heart muscle contraction.

2) What does the acronym *CLL* stand for?

Chronic lymphocytic leukaemia.

3) Who usually develops CLL – young, middle aged or older people?

Older. Usually greater than 50 years of age.

4) Which cells express CD8, helper T cells or cytotoxic T cells?

Cytotoxic T cells.

5) Name an inflammatory mediator responsible for vasodilation.

Nitric oxide, histamine and prostaglandins are inflammatory mediators responsible for vasodilation.

6) On a plain chest X-ray image, what structures make up the right border of the cardiac shadow?

Superior vena cava, right atrium and inferior vena cava.

7) What histological stain would you use to definitively identify mycobacterium tuberculosis?

Ziehl-Neelsen stain.

8) Name four examples of chronic inflammatory diseases.

Granulomatous disease, **autoimmune** disease and **longstanding infections**, repeated/**longstanding lesions** are categories of disease that lead to chronic inflammatory processes, for example:

Crohn's disease, Wegener's granulomatosis, Granuloma annulare etc.

Hashimoto's thyroiditis, Systemic lupus erythematosus, Polymyositis etc.

Tuberculosis, Chronic hepatitis, Helicobacter pylori gastritis, Chronic pyelonephritis, Chronic bronchitis etc.

Atherosclerosis; Ischaemic heart disease, Peripheral vascular disease, Ischaemic bowel etc.

Chronic pancreatitis, Post-surgical gastritis etc.

9) What is the plural of diverticulum?

Diverticula.

10) As a physician, what clinical features cause you to think of a diagnosis of an immunodeficiency disease?

Serious, persistent, unusual or recurrent infections ("SPUR").

11) What is the single word that means "enlarged liver"?

Hepatomegaly.

12) If blood collects in the pericardial space, what is it called, and what are the possible consequences?

A **haemopericardium** is the pathological term for blood in the pericardial space. If it becomes symptomatic it can cause life threatening **cardiac tamponade**.

13) Name two (endogenous) pyrogens responsible for initiating fever in the human body.

Interleukin 1

Interleukin 2

Interleukin 6

Tumour necrosis factor α

Prostaglandin E2

14) Which antibody is more useful in identifying an acute infection, IgM or IgG?

IgM (the first type of antibody to respond to the first challenge of an infective agent).

15) The median nerve supplies cutaneous sensation to which fingers?

The lateral 3.5 digits (thumb, index finger, middle finger and lateral aspect of ring finger).

16) Which is more worrying in the breast, a mobile palpable mass or a tethered palpable mass?

Tethered masses are more worrying because of the implied malignant infiltration of adjacent tissue.

17) Name three functions of macrophages.

Phagocytosis, cytokine production, tissue destruction and tissue repair and antigen presentation.

18) Name two systemic disorders that can occur as result of multiple myeloma.

Amyloidosis, hypercalcaemia, pathological fractures, anaemia, renal failure sequelae (e.g. hyperkalaemia) and relative immunosuppression leading to systemic infections.

19) Classically, with which groups of people do doctors communicate poorly?

Females.

The elderly.

Ethnic minorities.

Low socioeconomic status individuals.

Individuals with chronic illness.

Individuals with psychological symptoms.

20) Which nerve passes through the spiral groove of the humerus?

Radial nerve.

21) What effects does radial nerve damage have on the wrist?

Leads to wrist drop.

22) Numerically, what is the minimum BMI that is classed as obese?

30 kg/m^2

23) What does the acronym PTCA stand for?

Percutaneous transluminal coronary angioplasty.

24) Name four behaviours by doctors that are notorious for blocking patient disclosure.

Any of:

Not listening.

Depersonalization.

Explaining away a patient's distress as normal.

Attending to physical aspects of complaint only.

Jollying patients along.

Use of jargon.

25) Which disease produces the paraprotein called Bence-Jones protein?

Multiple myeloma.

26) If you were to give the following diseases one unifying classification, what would it be? Selective IgA deficiency, severe combined immunodeficiency, Di George syndrome and Bruton's X-linked agammaglobinaemia.

Primary immunodeficiency diseases.

27) Name two excised vessels that are used for coronary artery bypass grafts, CABG.

Any of:

Internal thoracic artery, long saphenous vein, radial artery and gastroepiploic artery.

28) Name the four lobes of the liver.

Right, left, caudate and quadrate.

29) What is myoglobin?

Monomeric oxygen binding protein found in skeletal muscle.

30) "A lower limit of 12g/dL in women and 14g/dL in men." To what is the speaker referring?

Haemoglobin concentration in the blood.

31) Give three examples of the health effects of poor communication between doctors and patients.

Patients are less likely to adhere to medical regimens. Patients are less likely to use healthcare services. Patients are less likely to use preventative healthcare. Patients are more likely to experience negative health outcomes.

32) Which spinal nerve root supplies all of the intrinsic muscles of the hand?

T1.

33) From which nerve roots does the brachial plexus originate?

C5, C6, C7, C8, T1.

34) From which white blood cell is the macrophage derived?

Monocyte.

35) What is the key genetic abnormality of Bruton's agammaglobinaemia?

It is an X-linked mutation in the Bruton's tyrosine kinase gene (Btk).

36) Are CD8 bearing cytotoxic T cells, MHC I or MHC II restricted?

MHC I restricted.

37) Does the sympathetic innervation of the heart decrease or increase the heart rate?

Increase.

38) On a plain chest X-ray image, which structures make up the right border of the cardiac shadow?

Aortic arch, pulmonary artery and left ventricle.

39) Is the Ghon focus in the apex of the lungs an example of acute or chronic inflammation?

Chronic.

40) After a shoulder dislocation, which nerve function should you test?

Axillary nerve e.g. by testing for abduction.

41) Which muscles mark the anterior border of the axilla?

Pectoralis major and pectoralis minor.

42) Which white blood cells are involved in chronic inflammation?

Macrophages, lymphocytes, plasma cells and eosinophils.

43) Which of the following cells are not directly involved in chronic inflammation; macrophages, plasma cells, lymphocytes or erythrocytes?

Erythrocytes.

44) What is the minimum length of time that must pass after a meal before the fasted state is reached?

4 hours.

45) Which chronic inflammatory cell has a "clock face" nucleus?

Plasma cell.

46) Name three disorders that an atopic individual would be more likely to experience.

Asthma, eczema, hayfever and allergies.

47) Explain the difference between a lymphoma and a leukaemia.

Leukaemia is a malignancy of blood cells that is primarily in the *circulatory* (haematogenous) *system* and *bone marrow*. In contrast lymphoma is a malignancy that starts in the lymph nodes or the lymphatic system and forms *tissue masses* earlier in its progression. Leukaemia usually manifests itself because the abnormal blood cells divide more rapidly and overwhelm the normal blood cells and prevent their function – hence relative immunosuppression and infections are common.

Session 16

1) Which of the following organs is not transplantable in humans; kidney, liver, heart, lung, spleen, small bowel?

2) Give two examples of drugs used to prevent tissue infection.

3) How many ATP molecules are hydrolyzed by the nicotinic acetylcholine receptor, when it is bound to acetylcholine?

4) What is secondary prevention?

5) What is a patent ductus arteriosus?

6) Give another name for epinephrine.

7) Which molecule must a G-protein bind to become active?

8) Histologically speaking, what type of cells are the effectors of tissue rejection?

9) What is another name for the follicular phase of the menstrual cycle?

10) When standing up straight and wearing trousers, contraction of which shoulder muscle is most important in allowing you to put your hand in the ipsilateral trouser pocket?

11) What type of receptor receives the neuromuscular signal from acetylcholine at skeletal muscle?

12) Where are you most likely to find coarctation the aorta?

13) What is the most likely cause of a clavicular fracture?

14) What is the difference between all allograft and an autograft?

15) Considering classical cardiovascular physiology, which two factors alone determine systemic blood pressure?

16) What type of collagen is in bone?

17) In relation to their adaptation to their illness, patients who believe that it is possible to combat illness are more likely to behave in what ways?

18) List four physiological fight or flight responses.

19) In immunology, what does the acronym TCR stand for?

20) What type of collagen is found in cartilage?

21) List the four components of the tetralogy of Fallot.

22) What does the acronym ECT stand for?

23) There are three general time courses for tissue rejection. What are they?

24) There are three major HLA types in humans. Name them.

25) What types of substances are fibronectin and laminin?

26) Name the type of collagen present in basement membranes.

27) Give another name for the *luteal* phase of the menstrual cycle.

28) Name two hormones/interleukins that can stimulate angiogenesis.

29) What is healing by primary intention?

30) What is the modern name for the internal mammary artery?

31) Define primary health care.

32) List five components of granulation tissue.

Session 16 Answers

1) Which of the following organs is not transplantable in humans; kidney, liver, heart, lung, spleen, small bowel?
Spleen.
2) Give two examples of drugs used to prevent tissue rejection.
Cyclosporin A, tacrolimus, prednisolone, azathioprine and mycophenolate.
3) How many ATP molecules are hydrolyzed by the nicotinic acetylcholine receptor, when it is bound to acetylcholine?
None. It is an ion channel that carries out facilitated transport not active transport.
4) What is secondary prevention?
Interventions aimed at detecting diseases early to delay or halt disease progression, e.g. screening.
5) What is a patent ductus arteriosus?
An open vascular connection between the proximal pulmonary artery and descending aorta - usually a congenital malformation.
6) Give another name for epinephrine.
Adrenaline.
7) Which molecule must a G-protein bind to become active?
After being triggered by the receptor-agonist complex the G protein must bind GTP to become active.
8) Histologically speaking, what type of cells are the effectors of tissue rejection?
T lymphocytes.
9) What is another name for the follicular phase of the menstrual cycle?
Proliferative phase.
10) When standing up straight and wearing trousers, contraction of which shoulder muscle is most important in allowing you to put your hand in the ipsilateral trouser pocket?
Contraction of the **deltoid muscle** allows you to raise your hand to

enter the pocket.

11) What type of receptor receives the neuromuscular signal from acetylcholine at skeletal muscle?

Nicotinic receptors.

12) Where are you most likely to find coarctation of the aorta?

At the junction of distal aortic arch and descending aorta below the origin of subclavian artery (distal to the insertion of the ductus arteriosus).

13) What is the most likely cause of a clavicular fracture?

Falling onto an outstretched hand.

14) What is the difference between an allograft and an autograft?

An allograft is donated tissue from same species but a different individual. An autograft is donated tissue to a recipient that is the same person as the donor.

15) Considering classical cardiovascular physiology, which two factors alone determine systemic blood pressure?

Cardiac output and total peripheral resistance.

16) What type of collagen is in bone?

Type 1 collagen.

17) In relation to their adaptation to their illness, patients who believe that it is possible to combat illness are more likely to behave in what ways?

Adapt to consequences of illness; attend rehabilitation programmes and adhere to treatment.

18) List four physiological fight or flight responses.

The fight or flight responses are numerous and include:

Dilation of the pupils.

Acceleration of the heart rate.

Acceleration of respiratory rate.

Diversion of blood to peripheral skeletal muscle.

Bronchodilation.

Reduced GI motility.

GI sphincter constriction.

Relaxation of the bladder.

Constriction of blood vessels of erectile tissue.

Dry mouth.

19) In immunology, what does the acronym TCR stand for?

T cell receptor.

20) What type of collagen is found in cartilage?

Type 2 collagen.

21) List the four components of the tetralogy of Fallot.

Subpulmonary stenosis.

Ventricular septal defect.

Overriding aorta.

Right ventricular hypertrophy.

22) What does the acronym ECT stand for?

Electroconvulsive therapy.

23) There are three general time courses for tissue rejection. What are they?

Hyperacute, acute and chronic.

24) There are three major HLA types in humans. Name them.

A, B and DR.

25) What types of substances are fibronectin and laminin?

Extracellular matrix glycoproteins.

26) Name the type of collagen present in basement membranes.

Type 4 collagen.

27) Give another name for the *luteal* phase of the menstrual cycle.

The secretory phase.

28) Name two hormones/interleukins that can stimulate angiogenesis.

Any two of:

VEGF (Vascular Endothelial Growth Factor)

bFGF (basic Fibroblast Growth Factor)

IL8 (Interleukin 8)

29) What is healing by primary intention?

Wound edges are brought to close apposition to allow direct growth across the deficit.

30) What is the modern name for the internal mammary artery?

Internal thoracic artery.

31) Define primary health care.

Health care initiatives aimed at maintaining or improving health among people who are currently free of symptoms (includes modification of risk factors).

32) List five components of granulation tissue.

Any five of:

Proliferating capillaries, erythrocytes, leukocytes, macrophages, other mixed inflammatory cells, myofibroblasts, fibroblasts, growth factors and other hormones, extracellular matrix proteins, antibodies, water and electrolytes.

Session 17

1) What does chemotaxis mean?

2) How many different types of emboli do you know? Name them.

3) Give 8 risk factors for venous thrombus formation.

4) List the CAGE questions.

5) What does the acronym IUCD stand for?

6) How long can a subdermal contraceptive implant last?

7) What does anhedonia mean?

8) Define virulence.

9) Where in the world would you be most likely to contract a schistosomiasis infection?

10) What is the lay term for the infestation caused by the species Taenia solium?

11) In general terms what is the possible fate of a thrombus?

12) Where in the human body would you expect to find the lines of Zahn?

13) Which two screening questions could you use to assess

nicotine dependence?

14) What is the term for a tissue denuded of mucosa?

15) What does a positive result to a gram stain look like?

16) A large embolus across the hilum of the right lung is noted. What is this likely to be called?

17) What does the acronym CBT stand for?

18) What is a paradoxical embolus?

19) What are the common techniques available for the diagnosis of a pulmonary embolus?

20) What does the acronym DIC mean and name one disease that can cause it.

21) Name the primary and classical ECG change associated with myocardial infarction.

22) Name the primary protein abnormality of haemophilia A.

23) What is tPA?

24) Name the protein which forms the net that captures platelets as part of blood clot formation.

25) Are staphylococcal and streptococcal bacteria gram negative or gram positive?

26) What triggers the intrinsic coagulation pathway?

27) In psychological terms, what class of disease is depression?

28) Psychodynamic therapy has two underlying assumptions. State them.

29) List the main actions of the biceps brachii muscle.

30) Name the type of cell from which platelets are derived.

31) What is Virchow's Triad?

32) What is the classic symptom triad of menopause?

33) Name the average age at which women go through the menopause.

34) What percentage of primary care visits are driven by psychological factors?

35) What does the acronym PDGF stand for?

36) Considering infectious disease, what is vertical transmission?

37) Excluding ganglia of the nervous system, what is a ganglion?

38) What is the modern name for pneumocystis carinii?

Session 17 Answers

1) What does chemotaxis mean?

Chemotaxis means migration towards a chemical stimulus. During an acute inflammatory response the neutrophils migrate towards an attractive chemical stimulus.

2) How many different types of emboli do you know? Name them.

Thromboembolism (arterial or venous), **amniotic fluid** embolism, **fat** embolism, **air** embolism, **nitrogen** embolism, **tumour** fragment embolism, **bone marrow** embolism, **foreign material** embolism, **mycotic** embolism (infective).

3) Give 8 risk factors for venous thrombus formation.

The classic risk factors are:

Relative immobility

Tissue damage - major surgery/bone fracture.

Malignancy.

Dehydration.

Disseminated intravascular coagulation.

Obesity.

Previous deep venous thrombosis.

Female.

Increasing age.

Risk factors for recurrent/unusual or young age of presentation:

Protein C deficiency.

Antithrombin III deficiency.

Protein S deficiency.

Antiphospholipid (antibody) syndrome.

Factor V Leiden mutation.

Increased Factor VIII activity.

Sticky platelet syndrome.

Homocystinaemia

Prothrombin G20210A mutation.

4) List the CAGE questions.

Have you ever felt that you should cut down on your drinking?

Have people annoyed you by criticizing your drinking?

Have you ever felt bad or guilty about your drinking?

Have you ever had a drink first thing in the morning to steady your nerves or to get rid of a hangover (eye opener)?

5) What does the acronym IUCD stand for?

Intrauterine contraceptive device.

6) How long can a subdermal contraceptive implant last?

The maximum duration is approximately five years.

7) What does anhedonia mean?

Loss of enjoyment of previously pleasurable activities.

8) Define virulence.

It is a quantitiative measure of pathogenicity of an organism. It indicates the likelihood of an organism causing disease.

9) Where in the world would you be most likely to contract a schistosomiasis infection?

The tropics.

10) What is the lay term for the species Taenia solium?

Tapeworm.

11) In general terms what is the possible fate of a thrombus?

Embolization, resolution (fibrinolysis), propagation, organisation and recanalization are possible fates of a thrombus.

12) Where in the human body would you expect to find the lines of Zahn?

Anywhere in the vasculature that has a mature blood clot.

13) Which two screening questions could you use to assess nicotine dependence?

How many cigarettes do you smoke per day? (>15 per day is considered a poor sign).

How soon after you wake up do you smoke your first cigarette? (Within 30 minutes of waking up is another poor sign).

14) What is the term for a tissue denuded of mucosa?

Ulcerated tissue or an **ulcer**.

15) What does a positive result to a gram stain look like?

Dark blue or black staining.

16) A large embolus across the hilum of the right lung is noted. What is this likely to be called?

Saddle embolus.

17) What does the acronym CBT stand for?

Cognitive behavioural therapy.

18) What is a paradoxical embolus?

A venous thromboembolus that has entered the arterial blood supply. Often this is via a ventricular septal defect (VSD) or atrial septal defect (ASD).

19) What are the common techniques available for the diagnosis of a pulmonary embolus?

V/Q scan (ventilation/perfusion scan) or CT pulmonary angiography.

20) What does the acronym DIC mean and name one disease that can cause it.

Disseminated intravascular coagulation. Pancreatitis and disseminated malignancy are classic causes of DIC.

21) Name the primary and classical ECG change associated with myocardial infarction.

ST elevation.

22) Name the protein abnormality of haemophilia A.

Clotting factor VIII.

23) What is tPA?

Tissue plasminogen activator.

24) Name the protein which forms the net that captures platelets as part of blood clot formation.

Fibrin.

25) Are staphylococcal and streptococcal bacteria gram negative or positive?

Gram positive.

26) What triggers the intrinsic coagulation pathway?

Subendothelial collagen.

27) In psychological terms, what class of disease is depression?

An affective disorder.

28) Psychodynamic therapy has two underlying assumptions. State them.

Current difficulties are based on childhood experiences.

The patient does not have a conscious awareness of their real motives or impulses.

29) List the main actions of the biceps brachii muscle.

Flexion and supination at the elbow joint.

30) Name the type of cell from which platelets are derived.

Megakaryocytes.

31) What is Virchow's Triad?

Abnormalities of the vessel wall.

Abnormalities of the blood flow.

Abnormalities of blood constituents.

32) What is the classic symptom triad of menopause?

Hot flushes, vaginal dryness, sweats.

33) Name the average age at which women go through the menopause.

51 years of age.

34) What percentage of primary care visits are driven by psychological factors?

60-70%.

35) What does the acronym PDGF stand for?

Platelet derived growth factor.

36) Considering infectious disease, what is vertical transmission?

Mother to child transmission (in utero or perinatally).

37) Excluding ganglia of the nervous system, what is a ganglion?

A synovial herniation.

38) What is the modern name for pneumocystis carinii?

Pneumocystis jiroveci.

Session 18

1) Which investigations would you use to diagnose polymyositis?

2) What is the first line antibiotic used for urinary tract infections?

3) Define stroke volume.

4) During exercise what is the CNS based control mechanism for the heart rate?

5) State a synonym for cancellous bone.

6) What is the probability of a Down's syndrome child being born to a 35 year old pregnant woman?

7) Name a class 1a antiarrhythmic drug.

8) Which two veins does the median cubital vein unite?

9) Briefly describe the pathogenesis of myasthenia gravis.

10) What is the lower threshold for a normal ejection fraction?

11) What is the more common name for *bone matrix* and what are its non-cellular components?

12) What hormone(s) are in the mini-contraceptive pill?

13) Define dislocation.

14) What does the acronym MSU mean? What does the acronym MCS stand for?

15) What does the acronym TORCH stand for?

16) Describe Starling's law of the heart.

17) What is the average female age at first pregnancy in the UK?

18) After a sore throat caused by Lancefield Group A β-haemolytic streptococci, which two immune related complications are notorious sequelae?

19) What features do you look for if you suspect compartment syndrome?

20) If you developed a skin infection, what would be the most likely bacterial cause?

21) List three drugs you could use to treat myasthenia gravis.

22) Which tumour is associated with myasthenia gravis?

23) What does the acronym MRSA stand for?

24) What is the Bainbridge reflex?

25) List six risk factors for atherogenesis.

26) Which cells can you find in bone?

27) What are the key components of the tunica media?

28) Explain the difference between iatrogenic and idiopathic.

29) What type of receptors mediate the sympathetic nervous system's signal at the pacemaker cells of the heart?

30) What is arteriolosclerosis?

31) Which of the following is not increased in response to fight or flight situations?
   Cardiac rate
   Blood pressure
   Respiratory rate
   Circulating catecholamines
   Cortisol
   Glycogen breakdown
   Gut function
   Peripheral diversion of blood

32) List five components of a mature atheroma.

Session 18 Answers

1) Which investigations would you use to diagnose polymyositis?
All three are helpful in reaching a diagnosis:
EMG (electromyography).
Blood creatine kinase concentration (muscle variant).
Biopsy and histopathology.
2) What is the first line antibiotic used for urinary tract infections?
Trimethoprim.
3) Define stroke volume.
The volume of blood ejected by the ventricle between the end of normal diastole and the normal systole.
4) During exercise what is the CNS based control mechanism for the heart rate?
The cerebral cortex stimulates the vasomotor centre that recruites the sympathetic nervous system that in turn has positive chronotropic and inotropic effects on the heart.
5) State a synonym for cancellous bone.
Trabecular bone or spongy bone.
6) What is the probability of a Down's syndrome child being born to a 35 year old pregnant woman?
Approximately 1/400.
7) Name a class 1a antiarrhythmic drug.
Quinidine, procainamide or disopyramide. This class are sodium channel blockers.
8) Which two veins does the median cubital vein unite?
Basilic vein and cephalic vein.
9) Briefly describe the pathogenesis of myasthenia gravis.
The disease is characterized by autoantibodies that bind and antagonize the nicotinic receptors at neuromuscular junctions.
10) What is the lower threshold for a normal ejection fraction?
55%.
11) What is the more common name for *bone matrix* and what are

its non-cellular components?

The bone matrix is osteoid. It includes type 1 collagen, proteoglycans, glycoproteins, alkaline phosphatase and water.

12) What hormone(s) are in the mini-contraceptive pill?

Low dose progestogen. It does not contain oestrogen.

13) Define dislocation.

Complete pathological separation of the articular surfaces of a joint.

14) What does the acronym MSU mean? What does the acronym MCS stand for?

MSU is mid stream urine.

MCS is microscopy, culture and sensitivity.

15) What does the acronym TORCH stand for?

Toxoplasmosis.

Other e.g. syphilis.

Rubella.

Cytomegalovirus.

Herpes simplex virus.

16) Describe Starling's law of the heart.

The stroke volume of the heart increases in response to an increase in the volume of blood filling the heart (the end diastolic volume). The force of cardiac muscle contraction is a function of the length of the muscle fibres.

17) What is the average female age at first pregnancy in the UK (2011)?

Approximately 31.

18) After a sore throat caused by Lancefield Group A β-haemolytic streptococci, which two immune related complications are notorious sequelae?

Poststreptococcal glomerulonephritis.

Infective endocarditis.

19) What features do you look for if you suspect compartment syndrome?

In the affected region look for the 5 Ps:

Pallor

Pulselessness

Paraesthaesia

Paralysis

Pain

20) If you developed a skin infection, what would be the most likely bacterial cause?

Staphylococcus aureus (streptococci are possible but less likely).

21) List three drugs you could use to manage myasthenia gravis.

Neostigmine, pyridostigmine, prednisolone, cyclosporine, mycophenalate or azathioprine.

22) Which tumour is associated with myasthenia gravis?

Thymoma.

23) What does the acronym MRSA stand for?

Methicillin resistant staphylococcus aureus.

24) What is the Bainbridge reflex?

Heart rate modification as a result of stimulation of atrial stretch receptors. (Increasing heart rate in response to an increase in central venous pressure).

25) List six risk factors for atherogenesis. Any six of:

Increasing age, male gender, family history, hyperlipidaemia, hypertension, cigarette smoking, diabetes mellitus, low HDL (high density lipoprotein) levels, physical inactivity, alcohol abuse, postmenopausal oestrogen deficiency.

26) Which cells can you find in bone?

Osteoblasts, osteocytes and osteoclasts.

Blood cells including lymphocytes, neutrophils and macrophages.

Adipocytes and stem cells/haemopoietic cells.

Pathological cells e.g. malignant cells.

27) What are the key components of the tunica media?

Smooth muscle and elastic tissue (containing elastin).

28) Explain the difference between iatrogenic and idiopathic.

An iatrogenic illness is one caused by doctors whereas an idiopathic illness has no identified cause.

29) What type of receptors mediate the sympathetic nervous system's signal at the pacemaker cells of the heart?

β_1 adrenergic receptors.

30) What is arteriolosclerosis?

Hypertension induced hypertrophy and fibrosis of arterioles.

31) Which of the following is not increased in response to fight or flight situations?

    Cardiac rate

    Blood pressure

    Respiratory rate

    Circulating catecholamines

    Cortisol

    Glycogen breakdown

    Gut function

    Peripheral diversion of blood

Gut function.

32) List five components of a mature atheroma.

Any five of:

Fibrous cap and fibroblasts.

Extracellular lipid. Cholesterol clefts.

Central necrosis.

Foamy macrophages.

T lymphocytes.

Smooth muscle cells.

Proliferating vessels.

Session 19

1) Define atrophy and name three possible causes.

2) Name the 3 major parts of the hip bone.

3) Name 3 types or classes of lower respiratory tract infections.

4) Outline the drug therapy for the treatment of a H. Pylori infection of the stomach.

5) What is the classic drug treatment for a chlamydia trachomatis infection?

6) How does verapamil work?

7) What is the classic drug treatment for herpes simplex?

8) What causes scabies?

9) How does digoxin work?

10) If a patient has a viral infection, which white blood cells preferentially dominate the response - neutrophils or lymphocytes?

11) Name five of the complications of diabetes mellitus that can occur during or after pregnancy.

12) Name two hyperglycaemic hormones.

13) Which muscarinic receptor mediates the negative chronotropic effect on the heart?

14) Why is pregnancy pro-thrombotic?

15) Which cells make β-hCG?

16) What vessels are in the umbilical cord?

17) Name three risk factors for an ectopic pregnancy.

18) How long after conception does implantation usually occur?

19) Why are there so many variants of the influenza A virus?

20) What is the name of each of the subunits of a normal G protein?

21) A 70 year old woman comes into an Accident and Emergency dept "off legs". What is at the top of your differential diagnosis?

22) What is placenta previa?

23) What does PET stand for?

24) Where in the body is the talus bone?

25) What happens to respiratory function during pregnancy?

26) The common peroneal nerve winds around the neck of which bone?

28) Which part of the femur is most prone to fracture?

29) Where in the body would you expect to find decidua?

Session 19 Answers

1) Define atrophy and name three possible causes.

Atrophy is shrinkage in cell size/tissue size by loss of cell substance. The causes may be any of the following:

Reduced workload

Loss of nerve supply.

Reduced blood supply.

Inadequate nutrition.

Loss of endocrine stimulation.

Aging.

2) Name the 3 major parts of the hip bone.

Ilium, ischium and pubis.

3) Name 3 types or classes of lower respiratory tract infections.

Pneumonia (lobar or broncho-)

Bronchitis

Bronchiolitis

Bronchiectasis

Empyema

Abscess

Chronic obstructive pulmonary disease

4) Outline the drug therapy for the treatment of a H. Pylori infection of the stomach.

Triple therapy comprises two different antibiotics and a proton pump inhibitor.

5) What is the classic drug treatment for a chlamydia trachomatis infection?

Doxycycline.

6) How does verapamil work?

It inhibits L-type calcium channels to reduce the intracellular calcium concentration and hence decrease the force of contraction through the decreased binding to troponin; the troponin-tropomyosin brake on muscle contraction is not released.

7) What is the classic drug treatment for herpes simplex?

Acyclovir.

8) What causes scabies?

Sarcoptes scabiei.

9) How does digoxin work?

It inhibits the sodium/potassium pump. This can have an antiarrhythmic effect by prolonging the repolarization phase of the cardiac action potential, thus increasing the duration of the cardiac action potential and so limiting the rate of the arrhythmia. The increase in intracellular sodium ion concentration stimulates the sodium/calcium exchanger in the cell membrane to increase the intracellular calcium concentration and so ultimately increasing the strength of myocyte contraction.

10) If a patient has a viral infection, which white blood cells preferentially dominate the response - neutrophils or lymphocytes?

Lymphocytes predominate in viral infections and neutrophils predominate in bacterial infections.

11) Name five of the complications of diabetes mellitus that can occur during or after pregnancy.

Any five of:

Miscarriage.

Stillbirth.

Preterm labour.

Foetal malformations.

Cerebral palsy in newborn.

Intrauterine growth retardation.

Macrosomia.

Malpresentation.

Shoulder dystocia and other birth injuries.

Cord prolapse.

Pre-eclamptic toxaemia

Poor maternal blood sugar control.

Development of gestational diabetes mellitus.

Neonatal hypoglycaemia.

Deterioration of renal function.

Deterioration of eye function.

Polyuria.

Polyhydramnios.

Polycythaemia rubra vera.

12) Name two hyperglycaemic hormones.

Any two of:

Adrenaline.

Noradrenaline.

Growth hormone.

Glucagon.

Human placental lactogen.

Cortisol.

Thyroid hormones (T3 and T4).

13) Which muscarinic receptor mediates the negative chronotropic effect on the heart?

M2.

14) Why is pregnancy pro-thrombotic?

Changes affecting all three parts of Virchow's triad apply:

Stasis – Venous compression by the pregnant uterus. Bedrest.

Hypercoagulability – Antiphospholipid syndrome. Dehydration.

Vascular damage – Varicose veins.

15) Which cells make β-hCG?

Syncytiotrophoblasts.

16) What vessels are in the umbilical cord?

Two umbilical arteries and one umbilical vein.

17) Name three risk factors for an ectopic pregnancy.

Any two of:

Previous pelvic infection (PID).

IUCD, Intrauterine contraceptive device.

Progesterone only pill.

Previous tubal surgery.

Previous abdominal surgery.

In vitro fertilization.

18) How long after conception does implantation usually occur?

7-9 days later.

19) Why are there so many variants of the influenza A virus?

Essentially because of *antigenic shift and drift*. There are multiple subtypes of neuraminidase and haemagglutinin proteins which thus allows for reassortment. Furthermore, the respective genes are subject to high mutation rates.

20) How many subunits are there on a complete G protein and what are they called?

There are three subunits. These are α, β and γ subunits.

21) A 70 year old woman comes into A+E "off legs". What is at the top of your differential diagnosis?

A urinary tract infection is a common cause of an older woman being suddenly unable to cope alone. The differential diagnosis includes hypothyroidism and dehydration.

22) What is placenta previa?

The placenta is wholly or partially sited in the lower uterine segment. It can be ruptured by the descending foetus and can lead to a catastrophic haemorrhage.

23) What does the acronym PET stand for?

Pre-eclamptic toxaemia; this is a medical/obstetric emergency that results in a critical hypertension. It is the second most common cause of pregnancy death directly due to the pregnancy.

24) Where in the body is the talus bone?

The ankle bone is the best answer. The foot would be acceptable.

25) What happens to respiratory function during pregnancy?

There is an increased respiratory rate.

There is an increased tidal volume.

There is an increased inspiratory volume.

There is a decreased residual volume.

There is a decreased expiratory reserve volume.

There is a reduction in functional residual capacity.

26) The common peroneal nerve winds around the neck of which bone?

Fibula.

28) Which part of the femur is most prone to fracture?

Neck of femur.

29) Where in the body would you expect to find decidua?

The pregnant uterus (superficial layers adjacent to the lumen).

Session 20

1) What are the actions of the sartorius muscle?

2) Does myocardial infarction cause ST depression or elevation?

3) What is the APGAR score?

4) Do the plantar interossei of the foot abduct or adduct?

5) What muscles are in the lateral compartment of the lower leg?

6) Name three potential complications of a rubella infection for a neonate.

7) What are the screening questions for body dysmorphic disorder?

8) Name the current first line antibiotic treatment against bacterial meningitis.

9) A patient presents with a community acquired pneumonia, what would be the most appropriate first line antibiotic?

10) On a death certificate what is the significance of 1a, 1b, 1c and 2?

11) What is left bundle branch block?

12) List the percentages of fat, lactose and water in mature human milk.

13) What is the Ferguson reflex?

14) Name the superficial muscles of the posterior compartment of the lower leg.

15) List five histological characteristics of malignant cells.

16) Explain what the word *rhabdomyosarcoma* means.

17) What is a pathological fracture?

18) Name the eponymously entitled test for a ruptured Achilles tendon.

19) What does the acronym DEXA stand for?

20) What is Paget's disease of the bone?

21) Name the three classic bone injuries that can occur from a fall onto an outstretched hand.

22) Describe the travel pathway of the cardiac action potential after sympathetic stimulation.

23) Name two ECG changes that you may see immediately after a myocardial infarction.

24) Name a recognized adverse effect on the kidney due to the use of vancomycin.

25) In what material are solid biopsy specimens usually embedded?

26) List four types of individuals/professions that you might see at a multidisciplinary team meeting.

27) Which organ produces oxytocin?

28) What percentages of all births are pre-term?

29) How accurate are the certificated causes of death as written by clinicians?

30) What is a frozen section?

31) Name the superficial muscles of the posterior compartment of the lower leg.

Session 20 Answers

1) What are the actions of the sartorius muscle?

Hip flexion, hip abduction, hip lateral rotation and knee flexion.

2) Does myocardial infarction cause ST depression or elevation?

Elevation.

3) What is the APGAR score?

A measure of neonatal physiological function immediately after birth. The features examined are tone, colour, pulse, respiration and responsiveness.

4) Do the plantar interossei of the foot abduct or adduct?

Adduct. (PAD, DAB).

5) What muscles are in the lateral compartment of the lower leg?

Fibularis longus (peroneus longus).

Fibularis brevis (peroneus brevis).

6) Name three potential complications of a rubella infection for a neonate.

Any three of:

Deafness, eye defects, congenital heart defects, microcephaly and low birthweight.

7) What are the screening questions for body dysmorphic disorder?

i) Are you worried about your appearance in any way?

ii) Does this concern preoccupy you? (Do you wish you could worry about it less?)

iii) What effect has this preoccupation with your appearance had on your life?

8) Name the current first line antibiotic treatment against bacterial meningitis.

Ceftriaxone (a cephalosporin).

9) A patient presents with a community acquired pneumonia, what would be the most appropriate first line antibiotic?

Amoxicillin (500mg tds po).

Penicllin would be an acceptable alternative.

10) On a death certificate what is the significance of 1a, 1b, 1c and 2?

1a – Immediate cause of death.

1b – Cause of 1a.

1c – Cause of 1b.

2 – Risk factors for the cause of death or significant comorbidity.

For example:

1a – Intracerebral haemorrhage.

1b – Cerebral metastases.

1c – Squamous cell carcinoma of the lung

2 – Smoking and a family history of lung malignancy.

11) What is left bundle branch block?

Inhibited transmission of the cardiac action potential to the left ventricle causing prolongation of the QRS complex.

12) List the percentages of fat, lactose and water in mature human milk.

Fat 2%, lactose 7% and water 90%.

13) What is the Ferguson reflex?

It is the positive feedback process whereby uterine contractions in labour cause a massive secretion of oxytocin. The sensory receptors in the cervix and vagina pass the signal from the contractions to the hypothalamus that then releases more oxytocin.

14) Name the superficial muscles of the posterior compartment of the lower leg.

Fibularis longus (peroneus longus).

Fibularis brevis (peroneus brevis).

15) List five histological characteristics of malignant cells.

Any five of:

Increased numbers of normal mitotic figures (normal mitoses).

Increased numbers of abnormal mitotic figures (abnormal mitoses).

Increased nuclear and cellular pleomorphism.

Increased nuclear to cytoplasmic ratios.

Invasion through basement membranes.

Prominent nucleoli.

Hyperchromatism.

Cytoplasmic mucin vacuolation (adenocarcinoma).

16) Explain what the word *rhabdomyosarcoma* means.

It is a malignant skeletal muscle tumour. The suffix –sarcoma implies that it is a "soft tissue" malignancy that is usually derived from mesodermal tissue.

17) What is a pathological fracture?

A fracture which occurs because a bone breaks in an area that is weakened by another disease process.

18) Name the eponymously entitled test for a ruptured Achilles tendon.

Simmond's test.

19) What does the acronym DEXA stand for?

Dual energy X-ray absorptiometry.

20) What is Paget's disease of the bone?

A chronic bone disorder that typically results in irregular enlarged deformed bones due to excessive breakdown and formation of bone tissue.

21) Name the three classic bone injuries that can occur from a fall onto an outstretched hand.

Scaphoid fracture.

Forearm (Colles) fracture.

Clavicular fracture.

22) Describe the travel pathway of the cardiac action potential after sympathetic stimulation.

i) Sinoatrial node activation.

ii) Atrial contraction and conduction (internodal pathway).

iii) Atrioventricular node activation.

iv) Conduction by Bundle of His.

v) Conduction along right and left bundle branches to Purkinje fibres.

vi) Ventricular conduction and contraction.

23) Name two ECG changes that you may see immediately after a myocardial infarction.

Any two of:

ST elevation.

T wave inversion.

Tachycardia.

(The pathological Q wave does not occur within the first few hours after an MI).

24) Name a recognized adverse effect on the kidney due to the use of vancomycin.

Renal failure due to acute interstitial nephritis or glomerulonephritis.

25) In what material are solid biopsy specimens usually embedded?

Histopathological assessment usually requires embedding of the tissue sample in wax.

26) List four types of individuals/professions that you might see at a multidisciplinary team meeting. Any four of the following:

Front line clinicians such as **surgeons** or **physicians**.

Nurses.

Pathologists.

Radiologists.

Administrators/personal assistants.

Medical students.

Trainee nurses.

27) Which organ produces oxytocin?

Posterior pituitary.

28) What percentages of all births are pre-term?

10%

29) How accurate are the certificated causes of death as written by clinicians?

Approximately 70%. 30% contain a major discrepancy between the post mortem findings and the clinician's death certificate information.

30) What is a frozen section?

A rapid histological assessment of a tissue sample that does not require embedding in paraffin wax but instead relies on the sample being frozen solid before assessment. A frozen section is a common intra-operative request to assess excision margins of malignancy, for example. However the procedure sacrifices diagnostic accuracy for speed.

31) Name the superficial muscles of the posterior compartment of the lower leg.

Gastrocnemius.

Soleus.

Plantaris.

Session 21

1) What are the initial treatments and definitive management of compartment syndrome?

2) List three methods for estimating gestational age.

3) Name two techniques for obtaining lung material suitable for the confirmation of infection.

4) Which has the higher affinity for oxygen – fetal haemoglobin or maternal haemoglobin?

5) When is the urinary system fully developed?

6) What does oligohydramnios mean?

7) What is the standard drug treatment for a pneumocystis jiroveci infection of the lungs?

8) What is reverse transcriptase and name a clinically important virus that carries this enzyme.

9) Considering pregnancy what does FH stand for?

10) Which two bones form the articulatory surface of the subtalar joint?

11) What do you understand by the phrase "in situ carcinoma"?

12) What is the average birth weight of a neonate? (2011, Western

World)

13) What are the fetal circulation shunts and which organs do they circumvent?

14) What does the acronym HIV stand for?

15) What name would you give a benign primary bone tumour derived from bone cells?

16) How is hepatitis A usually transmitted?

17) How can HIV be transmitted?

18) Does bradykinin cause vasoconstriction or vasodilation?

19) Describe Egan's four stage approach to counselling.

20) Name the vascular cancer that classically occurs after an HIV infection.

21) What is meconium?

22) If you accept that the development of a malignancy involves multistep oncogenesis, name four possible types of steps.

23) Describe the axes of motion of the ankle joint, subtalar joint and mid tarsal joint.

24) What is AZT and what is it used for?

25) Name five psychotherapies available in the NHS.

26) Considering cells and tissues, what is dysplasia?

27) What is watchful waiting as applied to the management of depression?

28) Which of the following is not a cause of chest pain?
Cardiac ischaemia
Pericarditis
Aortic dissection
Reflux oesophagitis
Abdominal aorta aneurysm
Trauma
Pulmonary embolism

Session 21 Answers

1) What are the initial treatments and definitive management of compartment syndrome?

a) Remove cast if present. Remove bandages if present. Examine skin.

b) The definitive management ultimately may require a fasciotomy to relieve the pressure.

2) List three methods for estimating gestational age.

i) Determining duration of pregnancy from the last menstrual period.

ii) Using fundal height to estimate foetal age.

iii) Using an ultrasound scan.

iv) Using biometric developmental criteria e.g. head circumference.

v) Using biophysical criteria e.g. presence of foetal breathing movements.

3) Name two techniques for obtaining lung material suitable for the confirmation of infection.

Obtaining sputum for culture and microscopy.

Bronchoalveolar lavage.

Lung biopsy.

4) Which has the higher affinity for oxygen – foetal haemoglobin or maternal haemoglobin?

Foetal.

5) When is the urinary system fully developed?

4-5 years of age.

6) What does oligohydramnios mean?

Too little amniotic fluid.

7) What is the standard drug treatment for a pneumocystis jiroveci infection of the lungs?

Co-trimoxazole.

8) What is reverse transcriptase and name a clinically important virus that carries this enzyme.

Reverse transcriptase catalyzes the conversion of viral RNA into its complementary DNA. The Human Immunodeficiency Virus (HIV) carries this enzyme.

9) Considering pregnancy what does FH stand for?

Fundal height.

10) Which two bones form the articulatory surface of the subtalar joint?

Talus and calcaneus.

11) What do you understand by the phrase "in situ carcinoma"?

A neoplasm with the cytological features of malignancy but no invasion (yet). There is no invasion through basement membranes.

12) What is the average birth weight of a neonate? (2011, Western World)

3.2 kg.

13) What are the fetal circulation shunts and which organs do they circumvent?

Ductus venosus – bypasses the liver.

Foramen ovale – bypasses the right ventricle and lungs by allowing the blood to pass from the inferior vena cava to the left atrium.

Ductus arteriosus – bypasses the lungs by allowing the blood to pass from the pulmonary artery to the aorta.

14) What does the acronym HIV stand for?

Human immunodeficiency virus.

15) What name would you give a benign primary bone tumour derived from bone cells?

Osteoma.

16) How is hepatitis A usually transmitted?

Faecal-oral transmission via unclean water.

17) How can HIV be transmitted?

Sexually (anal, vaginal or oral).

By vertical transmission.

By intravenous drug use.

By blood transfusion and blood products.

18) Does bradykinin cause vasoconstriction or vasodilation?

Vasodilation.

19) Describe Egan's three stage approach to counselling.

Stage One – The Present (understanding the current situation).

Stage Two – The Preferred (ideal/better life goal).

Stage Three – The Strategies (how to achieve the goal).

20) Name the vascular cancer that classically occurs after an HIV infection.

Kaposi's sarcoma.

21) What is meconium?

Foetal faeces.

22) If you accept that the development of a malignancy involves multistep oncogenesis, name four possible types of steps.

Oncogene activation.

Insensitivity to negative growth signals.

DNA repair defect.

Apoptotic failure.

Sustained angiogenesis.

Ability to invade and metastasize.

Limitless replicative potential.

23) Describe the axes of motion of the ankle joint and subtalar joint.

Ankle joint – plantar/dorsiflexion.

Subtalar joint – inversion/eversion, abduction/adduction.

24) What is AZT and what is it used for?

Zidovudine. This is a nucleoside analogue anti-retroviral drug used in the management of HIV infection.

25) Name five psychotherapies available in the NHS.

Egan's 3 or 4 stage approach.

Psychodynamic therapy.

Humanistic approach.

Systemic therapy.

Cognitive behavioural therapy.

26) Considering cells and tissues, what is dysplasia?

Cellular and tissue atypia that represents disordered differentiation and is premalignant. The atypia can range from mild to moderate to severe. Severe dysplasia is carcinoma in situ.

27) What is **watchful waiting** as applied to the management of depression?

Approximately 1/3 of patients with depression will recover within six weeks, so watchful waiting applies **reassurance** and **social facilitation without medical intervention**. Watchful waiting is more commonly applied to outpatients.

28) Which of the following is not a cause of chest pain?

Cardiac ischaemia

Pericarditis

Aortic dissection

Reflux oesophagitis

Abdominal aorta aneurysm

Trauma

Pulmonary embolism

Abdominal aortic aneurysms are not characterized by chest pain.

Session 22

1) Name the two organisms that commonly cause pelvic inflammatory disease (PID).

2) Immediately after a myocardial infarction, which is the more effective management, angioplasty or fibrinolysis?

3) A tumour produces a hormone that has systemic effects. What is the general phrase that describes this situation?

4) Which four drugs are used in the initial treatment of a tuberculosis infection?

5) Name four drugs that can be used to inhibit platelet function.

6) After three episodes of pelvic inflammatory disease what is the probability of subsequent infertility?

7) What is dyspareunia?

8) What are cadherins?

9) If a patient suffers a myocardial infarction, which ECG leads are most likely to show the ST elevation?

10) How do statins produce their positive clinical effect?

11) What is the classic cause of an ischial fracture?

12) What is the mortality risk associated with cardiac catheterization?

13) Name three growth factors important in facilitating angiogenesis.

14) What are the three key indications for the surgical treatment of stable angina?

15) Which cardiac event marker has troponin T largely replaced?

16) What are the eight steps that must occur for a malignant cell to move from its local environment to establish itself at its new metastatic site?

Session 22 Answers

1) Name the two organisms that commonly cause pelvic inflammatory disease (PID).

Chlamydia trachomatis.

Neisseria gonorrheae.

2) Immediately after a myocardial infarction, which is the more effective management, angioplasty or fibrinolysis?

Angioplasty.

3) A tumour produces a hormone that has systemic effects. What is the general phrase that describes this situation?

A **paraneoplastic syndrome**. For example the production of ADH by a primary small cell carcinoma of the lung is a paraneoplastic syndrome. This paraneoplastic syndrome causes the syndrome of inappropriate ADH production (SIADH), an intractable cause of hyponatraemia.

4) Which four drugs are used in the initial treatment of a tuberculosis infection?

RIPE is a helpful acronym:

Rifampicin, **I**soniazid, **P**yrazinamide and **E**thambutol.

(The continuation phase, which follows the initial treatment, usually uses rifampicin and isoniazid only).

5) Name four drugs that can be used to inhibit platelet function.

Aspirin.

Heparin.

Clopidogrel.

Ticlodipine.

Abciximab, Eptifibatide or Tirofiban. *These are Glycoprotein IIb/IIIa inhibitors.*

6) After three episodes of pelvic inflammatory disease what is the probability of subsequent infertility?

Approximately 50%.

7) What is dyspareunia?

Sexual intercourse that is painful.

8) What are cadherins?

Calcium dependent glycoproteins at the cell surface that mediate interactions between cells. Cadherins have reduced expression in many cancers, allowing the cells to be discohesive.

9) If a patient suffers a myocardial infarction, which ECG leads are most likely to show the ST elevation?

V1, V2 and V3.

10) How do statins produce their positive clinical effect?

They reduce cholesterol load by inhibiting the corresponding synthetic enzyme HMG-CoA reductase.

They stabilize coronary plaques.

11) What is the classic cause of an ischial fracture?

The major trauma that occurs in a road traffic accident.

12) What is the mortality risk associated with cardiac catheterization?

Approximately 1/1000.

13) Name three growth factors important in facilitating angiogenesis.

Any three of:

Vascular endothelial growth factor.

Basic fibroblast growth factor.

Angiopoietin 1.

Angiopoietin 2.

14) What are the three key indications for the surgical treatment of stable angina?

i) Symptoms continue despite adequate medical therapy.

ii) Lesions that are not suitable for angioplasty.

iii) The presence of prognostically significant coronary artery disease.

15) Which cardiac event marker has troponin T largely replaced?

CK-MB; the cardiac isozyme of creatine kinase.

16) What are the eight steps that must occur for a malignant cell to move from its local environment to establish itself at its new metastatic site?

Invasion of basement membrane.

Passage through the extracellular matrix.

Intravasation (into the blood vessel).

Immune interaction (avoidance of immune surveillance).

Platelet adhesion.

Adhesion to endothelium or basement membrane.

Extravasation.

Angiogenesis.

Session 23

1) What percentage of CIN3 cases progress to malignancy?

2) During surgery to remove a parotid gland tumour, which nerve is most likely to be accidentally damaged?

3) What effect does the action of aldosterone have on the body water balance and ion content?

4) What is bacterial endocarditis?

5) Which age group is more likely to get an endometrial carcinoma, under 50s or over 50s?

6) Why are beta blockers useful in heart failure?

7) Describe the clinically important effect of ultraviolet radiation on DNA.

8) Define SIRS. Under what circumstances does it most commonly occur?

9) What is scoliosis?

10) To which malignancy are individuals with coeliac disease particularly predisposed?

11) What are the three most important factors in grading a primary breast carcinoma?

12) To which malignancies does BRCA1 predispose?

13) Name five tumours of the female genital tract.

14) Why are loop diuretics used in heart failure?

15) Name three malignancies that EBV, the Epstein Barr virus, can cause.

16) Where would you find the nucleus pulposus?

17) Why is the use of digoxin problematic in heart failure?

18) What is an oncogene?

19) Which primary testicular malignancy is most likely to occur in 20-30 year old age group?

20) What are the two underlying processes that predispose to diabetic foot ulcers?

21) What is a fibroid?

22) To which malignancy are aniline dye workers predisposed?

23) Which quadrant of the breast is most likely to have a breast carcinoma and why?

24) Which malignancy is asbestos most notorious for causing?

25) What are rigors and why do they happen?

26) What is the name of the muscle of facial expression that surrounds the eye?

Session 23 Answers

1) What percentage of CIN3 cases progress to malignancy?
Approximately 30%.
2) During surgery to remove a parotid gland tumour, which nerve is most likely to be accidentally damaged?
Facial nerve.
3) What effect does the action of aldosterone have on the body water balance and ion content?
Sodium ions are retained, water is retained and potassium is lost.
4) What is bacterial endocarditis?
Infection of heart valves by bacteria.
5) Which age group is more likely to get an endometrial carcinoma, under 50s or over 50s?
Over 50s; the mean age at presentation is 55 years.
6) Why are beta blockers useful in heart failure?
They have a negative inotropic and negative chronotropic effect – making the heart more efficient. They reduce mortality.
7) Describe the clinically important effect of ultraviolet radiation on DNA.
The formation of pyrimidine dimers by the ultraviolet light increases the rate of mutations. The increased rate of mutations increases the risk of malignancy.
8) Define SIRS. Under what circumstances does it most commonly occur?
SIRS is the systemic inflammatory response syndrome. It is present when two or more of the following criteria are demonstrated in a patient:
a) A body temperature greater than 38°C or less than 36°C.
b) A heart rate greater than 90 beats per minute.
c) Tachypnoea greater than 20 breaths per minute or an arterial partial pressure of carbon dioxide of less than 32mmHg.

d) A white blood cell count of less than 4 x10^9 cells/L or greater than 12 x 10^9 cells/L.

SIRS is essentially systemic shock and its sequelae that often occur with overwhelming bacterial septicaemia.

9) What is scoliosis?

Lateral deviation and deformity of the spine.

10) To which malignancy are individuals with coeliac disease particularly predisposed?

Lymphoma.

11) What are the three most important factors in grading a primary breast carcinoma?

Tubules; presence and quantity.

Extent of **pleomorphism**.

Presence and quantity of **mitoses**.

12) To which malignancies does BRCA1 predispose?

Breast carcinoma.

Ovarian carcinoma.

Uterine carcinoma.

Cervical carcinoma.

Prostate carcinoma.

Pancreatic carcinoma.

Colon carcinoma.

13) Name five tumours of the female genital tract.

Tumours comprise benign and malignant neoplasms. Furthermore the malignant neoplasms may be primary or secondary. Therefore *some* of the common neoplasms are listed:

Cervical squamous cell carcinoma.

Cervical adenocarcinoma.

Leiomyoma (fibroids). Leiomyosarcoma.

Endometrial adenocarcinoma. Endometrial stromal tumour.

Ovarian carcinoma. Borderline ovarian tumour.

Metastases such as the Krukenberg tumour.

Metastatic adenocarcinoma from any other intestinal site.

14) Why are loop diuretics used in heart failure?

To control fluid overload; to remove fluid that is causing pulmonary oedema and compromising the cardiovascular system.

15) Name three malignancies that EBV, the Epstein Barr virus, can cause.

Burkitt lymphoma. Hodgkin's lymphoma. Nasopharyngeal carcinoma. Lymphomas of the elderly. Post-transplant lymphoproliferative disorders.

16) Where would you find the nucleus pulposus?

In the center of the intervertebral disc.

17) Why is the use of digoxin problematic in heart failure?

It has a narrow therapeutic range. (As a general observation it should be noted that all antiarrhythmics used at high concentrations can cause arrhythmias).

18) What is an oncogene?

An activated proto-oncogene; the latter is a control molecule of normal growth that has mutated to become overexpressed or overactive, essentially sending a permanent growth signal. This predisposes to malignancy.

19) Which primary testicular malignancy is most likely to occur in 20-30 year old age group?

Teratoma.

20) What are the two underlying processes that predispose to diabetic foot ulcers?

Ischaemic damage to the tissues because of occluding atherosclerosis.

Neuropathic damage that decreases the sensation of trauma to the foot.

21) What is a fibroid?

A leiomyoma; this is a benign smooth muscle tumour. It is usually found in the female reproductive tract, with the uterus being the most favoured location.

22) To which malignancy are aniline dye workers predisposed?

Bladder malignancy; transitional cell carcinoma.

23) Which quadrant of the breast is most likely to have a breast carcinoma and why?

The upper outer quadrant. The long axillary tail means that there is more breast mass in this quadrant.

24) Which malignancy is asbestos most notorious for causing?

Mesothelioma.

25) What are rigors and why do they happen?

The pyrogens produced by the septicaemia alter the temperature set point in the hypothalamus ("resetting the thermostat"). The result is that the thermoregulatory centre believes that the body is cooler than its real temperature. Reflexes are initiated to generate heat, one of which causes muscle spasms – these are rigors.

26) What is the name of the muscle of facial expression that surrounds the eye?

Orbicularis oculi.

Session 24

1) Which of the following is not a notifiable disease? Anthrax, mumps, cholera, meningitis, sarcoidosis, leptospirosis or tuberculosis.

2) What is a Dukes B adenocarcinoma?

3) Typically, does radial or ulnar deviation occur in rheumatoid arthritis?

4) What is anuria?

5) Where in the body is the transformation zone?

6) Which types of epithelia border the transformation zone?

7) Name the two major types of psychological medicine treatment strategies.

8) Considering the management of breast disease, what is the triple approach?

9) Approximately what percentage of primary breast malignancies are HER2 positive?

10) Name four different types of shock.

11) To what type of cell is the word *dyskaryosis* usually applied?

12) What is a collapsing pulse?

13) Generally, what happens to neurons in the brain as part of the addiction process?

14) Which disease can cause a boutonniere deformity?

15) Are oestrogen receptors found predominantly in the cytoplasm or the nucleus?

16) What is the commonest autoimmune disease that might cause a 35 year old man to have a bamboo spine?

17) Explain the phrase "contrecoup injury."

18) What is a Duke's D adenocarcinoma?

19) Name the first line disease modifying drug in the management of rheumatoid arthritis.

20) What is a carcinoid tumour?

21) In public health, what does the acronym HPA mean?

22) Which is a more specific test for rheumatoid arthritis, rheumatoid factor or anti-cyclic citrullinated peptide?

23) What does the acronym MCCD mean?

24) A 25 year old woman has a small mobile untethered breast mass with a smooth surface. What is the most likely diagnosis?

Session 24 Answers

1) Which of the following is not a notifiable disease? Anthrax, mumps, cholera, meningitis, sarcoidosis, leptospirosis or tuberculosis.

Sarcoidosis is believed to be an autoimmune disease – it is not a notifiable disease. It is not infectious.

2) What is a Dukes B adenocarcinoma?

It is a colorectal adenocarcinoma that has extended through the muscularis propria. The five year survival rate is approximately 70%.

3) Typically, does radial or ulnar deviation occur in rheumatoid arthritis?

Ulnar deviation.

4) What is anuria?

Anuria means no urine output. If the kidney experiences prolonged anuria then irreversible renal damage will occur.

5) Where in the body is the transformation zone?

In the cervix.

6) Which types of epithelia border the transformation zone?

Non-keratinized stratified squamous equilibrium and simple columnar epithelium (as part of glandular mucosa).

7) Name the two major types of psychological medicine treatment strategies.

i) Brief psychological intervention.

ii) Psychotherapy.

8) Considering the management of breast disease, what is the triple approach?

Assessment by clinical signs and symptoms.

Radiological assessment.

Histological/cytological assessment.

9) Approximately what percentage of primary breast malignancies are HER2 positive?

20%.

10) Name four different types of shock.

Any four of:

Hypovolaemic shock.

Cardiogenic shock.

Septic shock.

Anaphylactic shock.

Mechanical shock.

11) To what type of cell is the word *dyskaryosis* usually applied?

Squamous cells of the cervix.

12) What is a collapsing pulse?

This is the sign that occurs because of aortic valve incompetence. It can also cause "pistol shot" femorals, head nodding and flushing of fingernails; they are the result of the dramatic drop in aortic blood pressure as the aortic valve collapses.

13) Generally, what happens to neurons in the brain as part of the addiction process?

The neurons adapt to the repeated drug exposure such that normal functioning only occurs in the presence of the drug.

14) Which disease can cause a boutonniere deformity?

Rheumatoid arthritis.

15) Are oestrogen receptors found predominantly in the cytoplasm or the nucleus?

In the nucleus. Oestrogen initiates and coordinates its effects by acting at a transcriptional level.

16) What is the commonest autoimmune disease that might cause a 35 year old man to have a bamboo spine?

Ankylosing spondylitis.

17) Explain the phrase "contrecoup injury."

A contrecoup injury is a deceleration injury on the side of the brain that is *opposite* to the initial site of the traumatic impact.

18) What is a Duke's D adenocarcinoma?

A colorectal adenocarcinoma that has produced distant metastases (i.e. to organs beyond the colon and rectum).

19) Name the first line disease modifying drug in the management of rheumatoid arthritis.

Methotrexate.

20) What is a carcinoid tumour?

A neuroendocrine tumour that is capable of producing serotonin (5-HT). It is often situated in the lungs or small intestine. It can cause a characteristic set of systemic effects that is called carcinoid syndrome.

21) In public health, what does the acronym HPA mean?

Health Protection Agency.

22) Which is a more specific test for rheumatoid arthritis, rheumatoid factor or anti-cyclic citrullinated peptide?

Anti-CCP. It is approximately 95% specific.

23) What does the acronym MCCD mean?

Medical certificate cause of death.

24) A 25 year old woman has a small mobile untethered breast mass with a smooth surface. What is the most likely diagnosis?

Fibroadenoma.

Session 25

1) What is the half-life of a drug?

2) What is patient adherence?

3) What effect does alcohol have on warfarin action?

4) What class of antibiotic is gentamicin and is it more effective against gram positive or gram negative bacteria?

5) Which leads to greater bioavailability, oral or intravenous drug administration?

6) Why is using creatinine clearance as a measure of GFR, glomerular filtration rate problematic; what assumptions are made?

7) The study of the *mechanism of action and effect of a drug* refers to which branch of pharmacology?

8) Where in a neurone does the most metabolic activity occur?

9) What is the minimum amount of fluid that a normal person must drink daily?

10) Into what class of antibiotics do penicillin and cephalosporins fall?

11) What does the acronym MRSA stand for?

12) In its anti-diuretic role in the kidney to what receptor does ADH bind?

13) Assuming a drug has successfully entered the circulatory system, name four factors that can affect tissue distribution.

14) What is malignant hyperthermia?

15) Which antibiotic should you chose to treat a serious pseudomonas infection?

16) What is the underlying lesion in diabetes insipidus?

17) Considering the metabolism of drugs in the liver, what are phase 1 and phase 2 reactions?

18) Give another name for Coumadin or coumarin.

19) What volume of blood is there in an average 70kg man?

20) What does the acronym SIADH stand for?

21) List four signs of hypervolaemia.

22) Describe the first-pass effect.

23) What is co-amoxiclav?

24) Which of the following is not a route of drug administration?

25) What common effect do salbutamol, salmeterol and terbutaline have on the lungs?

26) How does clavulanic acid work?

27) Name the three fundamental processes for which the nephron is responsible.

Session 25 Answers

1) What is the half-life of a drug?

The half-life of a drug is the time it takes for the total quantity of the drug in the body to drop by 50%.

2) Explain the phrase "patient adherence."

Patient adherence is synonymous with patient compliance; it is the willingness of the patient to follow the offered medical management.

3) What effect does alcohol have on warfarin action?

Alcohol potentiates/increases the action of warfarin by inhibiting warfarin breakdown (catabolism).

4) What class of antibiotic is gentamicin and is it more effective against gram positive or gram negative bacteria?

Gentamicin is an aminoglycoside that is most effective against Gram negative bacteria.

5) Which leads to greater bioavailability, oral or intravenous drug administration?

Intravenous administration leads to greater bioavailability. The definition of bioavailability refers to the proportion of the administered drug that reaches the venous circulation. By definition this is 100% for intravenous administration.

6) Why is using creatinine clearance as a measure of GFR, problematic; what assumptions are made?

Using creatinine clearance to estimate the glomerular filtration rate assumes that creatinine is freely filtered and not reabsorbed or secreted.

It assumes that creatinine production is constant and does not change over time.

It assumes that the measurement of creatinine is accurate and reproducible across laboratories.

7) The study of the *mechanism of action and effect of a drug* refers to which branch of pharmacology?

Pharmacodynamics. Namely, what the drug does to the body.

8) Where in a neurone does the most metabolic activity occur?

The body or *soma* - probably because of the presence of the nucleus.

9) What is the minimum amount of fluid that a normal person must drink daily?

400ml

10) Into what class of antibiotics do penicillin and cephalosporins fall?

These antibiotics have beta lactam rings and are termed beta lactams.

11) What does the acronym MRSA stand for?

Methicillin Resistant Staphylococcus Aureus.

12) In its anti-diuretic role in the kidney to what receptor does ADH bind?

The V_2 receptor. (ADH = Vasopressin)

13) Assuming a drug has successfully entered the circulatory system, name four factors that can affect tissue distribution.

Plasma protein binding.

Regional blood flow.

Lipid solubility.

Active transport.

Drug interactions.

Other concurrent disease processes.

14) What is malignant hyperthermia?

Malignant hyperthermia is an uncontrolled acute rise in body temperature secondary to general anaesthetic administration.

15) Which antibiotic should you chose to treat a serious pseudomonas infection?

Imipenem, meropenem or entapenem.

16) What is the underlying lesion in diabetes insipidus?

Diabetes insipidus occurs as a failure of ADH action, either through impaired formation, impaired secretion or failure of action at the kidney.

17) Considering the metabolism of drugs in the liver, what are phase 1 and phase 2 reactions?

Phase 1 reactions include oxidation, reduction and hydrolysis. These reactions may alter the activity of the drug or alter its water solubility.

Phase 2 reactions are generally conjugation reactions that usually facilitate clearance or excretion by the kidney/bile.

18) Give another name for Coumadin or coumarin.

Warfarin.

19) What volume of blood is there in an average 70kg man?

Approximately 4.5 litres.

20) What does the acronym SIADH stand for?

Syndrome of inappropriate anti diuretic hormone secretion.

21) List four signs of hypervolaemia.

Hypervolaemia implies too much fluid in the circulatory system. Signs may include:

Generalized oedema, sacral oedema, ankle swelling.

Pulmonary crepitations/oedema.

Dyspnoea, tachypnea.

Raised JVP.

Weight gain.

Hypertension.

22) Describe the first-pass effect.

The first-pass effect is the extent of metabolism (catabolism) of a drug before it enters the systemic circulation. It is often expressed as the percentage of the administered drug that reaches the systemic circulation.

23) What is co-amoxiclav?

An antibiotic that is a mixture of amoxicillin and clavulanic acid.

24) Which of the following is not a route of drug administration? Sublingual, buccal, nasal mucosal, transkeratinous, eye (optic), transdermal and inhalation.

"Transkeratinous" is not a route of drug administration.

25) What common effect do salbutamol, salmeterol and terbutaline have on the lungs?

They cause bronchodilation.

26) How does clavulanic acid work?

Clavulinic acid is a beta lactamase inhibitor. It binds to the bacterial beta-lactamase enzyme, thus preventing the bacteria from expressing resistance to beta lactam antibiotics.

27) Name the three fundamental processes for which the nephron is responsible.

Glomerular filtration/ultrafiltration.

Tubular secretion.

Tubular reabsorption.

Session 26

1) What are the disadvantages of carrying out an IVU (or IVP)?

2) Name a potassium sparing diuretic.

3) What are normal Babinksi reflexes and a positive Babinski sign?

4) What is the I$_f$ current?

5) To what diseases or disorders does *horseshoe kidney* predispose?

6) What is the name of the epithelium that lines the inner surfaces of the kidney and bladder?

7) What are the classically observed signs and symptoms of lower motor neuron lesions?

8) What is Conn's syndrome?

9) What effect do calcium channel blockers have on the heart?

10) What are Golgi tendon organs and what do they do?

11) There are seven modern drug class options for the management of hypertension. Name them.

12) Name two nitrovasodilators used in the management of angina.

13) Name three congenital abnormalities of kidney placement.

14) pH = pK + log$_{10}$ [Base]/[Acid] What is the name of this equation?

15) Explain what is meant by chorea.

16) Considering neurology, what is a reflex?

17) In managing a patient with hypertension who does not wish to take drugs, what other options can be tried?

18) What structures would you expect to see if you looked at the renal medulla under the light microscope?

19) Demonstrate the standard test for dysdiadochokinesia. What does a negative result probably indicate?

20) Where in the body would you expect to find a phaeochromocytoma?

21) Does nitric oxide increase cAMP or cGMP levels in target cells?

22) What are foot processes?

23) What is a KUB X-ray image and what is its significance?

24) Which is more soluble in blood, CO_2 or O_2?

25) What % of an average human's nephrons develop after birth?

26) Name the movement problems that can occur in Parkinson's disease.

27) What is the upper threshold for normal blood pressure?

28) What is the name for part of the nephron immediately distal to the loop of Henlé?

29) Does urinary system develop from ectoderm, mesoderm or endoderm?

30) How many O_2 molecules can one haemoglobin molecule hold?

Session 26 Answers

1) What are the disadvantages of carrying out an IVU (or IVP)?

Either technique exposes the patient to a radiation dose.

The contrast agent used is nephrotoxic or may cause patient sensitivity.

These techniques often have difficulty distinguishing between solid and cystic lesions or between benign and malignant neoplasms.

These techniques require normal renal function to be most effective.

2) Name a potassium sparing diuretic.

Amiloride and spironolactone are common potassium sparing diuretics.

3) What are normal Babinksi reflexes and a positive Babinski sign?

A normal Babinski reflex (i.e. no neurological lesion) causes a plantarflexion response.

A positive Babinski sign occurs when the ipsilateral toes dorsiflex as the Babinski reflex is triggered.

4) What is the I_f current?

The I_f current ("funny" current) is a cardiac pacemaker current that is activated by hyperpolarization. The current occurs because of the inward movement of Na^+ and K^+ ions through the cell membrane.

5) To what diseases or disorders does *horseshoe kidney* predispose?

Structural abnormalities of the urological tract are more likely to cause urinary stasis, which in turn makes **infection** more likely. The infection can then act as a nidus for **renal stone** formation.

6) What is the name of the epithelium that lines the inner surfaces of the kidney and bladder?

Urothelium which is **transitional cell epithelium**.

7) What are the classically observed signs and symptoms of lower motor neuron lesions?

Flaccid paralysis.

Absent or diminished spinal reflexes.

Muscular atrophy.

Muscular fasciculation.

8) What is Conn's syndrome?

Primary hyperaldosteronism. Adrenal hyperplasia is the commonest cause. Functioning adrenal adenomas are uncommon causes.

9) What effect do calcium channel blockers have on the heart?

They can decrease the heart rate (negative chronotropic effects). They can decrease the force of cardiac contraction (negative inotropic effects).

10) What are Golgi tendon organs and what do they do?

These organs are receptors sensitive to increases in muscle tension ("inverse stretch receptors"). Thus they can protect muscle and tendons from damage by preventing hyperextension or hyperflexion.

11) There are seven modern drug class options for the management of hypertension. Name them.

Diuretics.

α (adrenergic) blockers.

β (adrenergic) blockers.

Calcium channel blockers.

ACE inhibitors.

Renin inhibitors.

Angiotensin II receptor blockers.

12) Name two nitrovasodilators used in the management of angina.

Any two of:

Glyceryl trinitrate.

Isosorbide mononitrate.

Isosorbide binitrate (or isosorbide dinitrate).

13) Name three congenital abnormalities of kidney placement.

Pelvic kidneys.

Thoracic kidneys.

Crossed ectopia (both kidneys on the same side).

14) pH = pK + log$_{10}$ [Base]/[Acid] What is the name of this equation?

Henderson-Hasselbalch equation.

15) Explain what is meant by chorea.

"*Dance-like*" movements. An often quoted example is Huntington's chorea; this neurodegenerative disease is characterized by abnormal involuntary writhing movements.

16) Considering neurology, what is a reflex?

A simple stereotyped motor response to a specific type of stimulus.

17) In managing a patient with hypertension who does not wish to take drugs, what other options can be tried?

Lifestyle changes can be suggested to:

Control bodyweight.

Increase exercise.

Reduce salt intake.

Increase potassium intake.

Control alcohol intake.

18) What structures would you expect to see if you looked at the renal medulla under the light microscope?

Loop of Henle.

Capillaries.

Collecting ducts.

Interstitium.

19) Demonstrate the standard test for dysdiadochokinesia. What does a negative result probably indicate?

A cerebellar lesion.

20) Where in the body would find a phaeochromocytoma?

In the adrenal gland usually. Only 10% are extra-adrenal.

21) Does nitric oxide increase cAMP or cGMP levels in target cells?

cGMP is synthesized because of the nitric oxide stimulus.

22) What are foot processes?

Foot processes are podocytes. They border the urinary space aspect of the basement membrane. Foot processes are diminished ("effaced") in minimal change nephropathy – the commonest cause of nephrotic syndrome in children.

23) What is a KUB X-ray image and what is its significance?

KUB is a plain X-ray radiograph that images the kidneys, ureters and bladder. It is a quick, simple, readily available technique. It is useful for detecting renal stones and for noting gases.

24) Which is more soluble in blood, CO_2 or O_2?

Carbon dioxide, largely because carbonic anhydrase facilitates the equilibrium with dissolved bicarbonate.

25) What % of an average human's nephrons develop after birth?

0%. The nephrons develop before birth.

26) Name the movement problems that can occur in Parkinson's disease.

Akinesia – no movement.

Bradykinesia – slow movement.

Cog-wheel rigidity.

Tremor at rest.

27) What is the upper threshold for normal blood pressure?

A diastolic blood pressure greater than 90 mmHg is generally accepted to indicate systemic hypertension. Hence 139/89 is the upper threshold of normal systemic blood pressure.

28) What is the name for part of the nephron immediately distal to the loop of Henlé?

Distal convoluted tubule.

29) Does urinary system develop from ectoderm, mesoderm or endoderm?

It develops from intermediate **mesoderm**.

30) How many O_2 molecules can one haemoglobin molecule hold?

Four.

Session 27

1) What are Beauchamp and Childress's four principles of medical ethics?

2) How does the circulating concentration of K^+ change in either acute or chronic renal failure?

3) What are pedicels?

4) What is diabetes insipidus and why does it occur?

5) Where is the best perfused part of a normal lung?

6) What are the components of the brainstem?

7) Generally speaking under what circumstances can a competent child's refusal to consent be overridden?

8) What is the physiologically most important active form of angiotensin I?

10) What is the term for a significant pO_2 decrease in the body?

11) In what form is most CO_2 carried in the blood?

12) What does the letter V stand for in respiratory physiology?

13) Under what conditions is an advance refusal of treatment valid and acceptable?

14) What is a urachal fistula?

15) What is anuria and why is it particularly important to a junior doctor?

16) Which part of the body has the chemoreceptors that respond to pH changes in the patient's CSF?

17) What is microcephaly?

18) What type of consent is given when a patient offers their arm for a blood sample (venesection)?

19) In what direction does the blood flow in relation to the kidney in an afferent arteriole?

20) List four functions of the renal mesangium.

21) What is the term for a significant increase in pCO_2 in the body?

22) Where in the circulatory system would you find carbonic anhydrase?

23) Which artery is classically ruptured in an epidural haematoma?

24) Name the type of collagen present in the glomerular basement membrane.

25) What does the acronym CKD stand for?

26) Considering respiratory physiology, what does the letter Q stand for?

27) What is hydrocephalus?

28) What are the four components that make up the capacity to give consent?

29) Describe the Nernst equation.

30) Are glomeruli found in the renal medulla or cortex?

31) Where is the respiratory centre?

32) Why is O_2 unloading at active tissues so effective?

33) What is the name of the collateral circulation at the base of the brain?

34) What is deontology?

Session 27 Answers

1) What are Beauchamp and Childress's four principles of medical ethics?

Autonomy.

Justice.

Beneficence.

Non-maleficence.

2) How does the circulating concentration of K^+ change in either acute or chronic renal failure?

The plasma K^+ concentration rises in both acute and chronic renal failure and can result in hyperkalaemia.

3) What are pedicels?

Pedicels exist in the kidney as part of the nephron; they are the foot processes of podocytes.

4) What is diabetes insipidus and why does it occur?

This is a disease that occurs because of a failure of ADH function – leading to a diuresis (polyuria), increased thirst and increased water consumption. The disease is divided into two types – Central or Nephrogenic. Central type arises out of impaired vasopressin production. Nephrogenic type arises because of an unresponsiveness of the kidneys to ADH.

5) Where is the best perfused part of a normal lung?

The base of the lung.

6) What are the components of the brainstem?

Midbrain, pons and medulla.

7) Generally speaking under what circumstances can a competent child's refusal to consent be overridden?

"If valid consent is obtained from someone with the authority to consent." Typically this is usually a parent.

8) What is the physiologically active form of angiotensin I?

Angiotensin I has no direct physiological action alone. Is active form is effectively *angiotensin II*.

10) What is the term for a significant pO_2 decrease in the body?
Hypoxia.

11) In what form is most CO_2 carried in the blood?
Bicarbonate ions in the plasma.

12) What does the letter V stand for in respiratory physiology?
Ventilation e.g. V/Q scan for a pulmonary embolus.

13) Under what conditions is an advance refusal of treatment valid and acceptable?
The following conditions should be satisfied:

- The individual should be over 18.
- The individual must have had capacity at the time of writing.
- The advance refusal must be valid (e.g. not subsequently withdrawn).
- The advance refusal must be applicable to the treatment in question.
- The advance refusal should be in writing, signed and witnessed.

14) What is a urachal fistula?
A urachal fistula occurs as a result of the failure of obliteration of the allantois – allowing the bladder to communicate with the exterior. This fistula allows dribbling of urine to the skin surface at the umbilicus.

15) What is anuria and why is it particularly important to a junior doctor? Anuria is the failure of the kidneys to produce any urine. **Prolonged anuria can cause permanent renal damage.** Routine monitoring of renal function is usually the job of the junior doctor.

16) Which part of the body has the chemoreceptors that respond to pH changes in the patient's CSF?
Medulla and pons.

17) What is microcephaly?

A congenital abnormality in which there is underdevelopment of the brain.

18) What type of consent is given when a patient offers their arm for a blood sample (venesection)?

Implied.

19) In what direction does the blood flow in relation to the kidney in an afferent arteriole?

Towards the kidney.

20) List four functions of the renal mesangium.

Phagocytosis.

Synthesis of mesangial matrix.

Secretion of prostaglandins and endothelins.

Control of glomerular filtration.

Support of capillaries.

Turnover of basal lamina.

Regulation of blood flow.

Response to angiotensin II.

21) What is the term for a significant increase in pCO_2 in the body?

Hypercapnia.

22) Where in the circulatory system would you find carbonic anhydrase?

In erythrocytes (red blood cells).

23) Which artery is classically ruptured in an epidural haematoma?

The middle meningeal artery.

24) Name the type of collagen present in the glomerular basement membrane.

Type 4 collagen.

25) What does the acronym CKD stand for?

Chronic kidney disease.

26) Considering respiratory physiology, what does the letter Q stand for?

Perfusion.

27) What is hydrocephalus?

An abnormal and excessive accumulation of cerebrospinal fluid in its CNS circulation. This can lead to raised intracranial pressure and an enlarged head.

28) What are the four components that make up the capacity to give consent?

- Ability to understand the relevant information.
- Ability to retain that information.
- Ability to weigh/assess the information as part of the decision making process.
- Ability to communicate the decision.

29) Describe the Nernst equation.

$E_{ion} = (RT/zF)\lg[Ion]_{out}/[Ion]_{in}$

where R = Universal gas constant, T = Absolute temperature, z = number of electrons in moles and F = Faraday constant.

E_{ion} is the membrane potential (electromotive force) due to the ions of interest and $[Ion]_{out}$ is the extracellular ion concentration of interest and $[Ion]_{in}$ is the intracellular ion concentration.

30) Are glomeruli found in the renal medulla or cortex?

The cortex. The collecting tubules are a feature of the medulla.

31) Where is the respiratory centre?

In the pons and the medulla.

32) Why is O_2 unloading at active tissues so effective?

The features at active tissues which facilitate O_2 unloading from haemoglobin are:

Increased carbon dioxide concentration and lower pH.

Increased 2,3-BPG concentration.

Higher temperature.

They tend to stabilize the deoxygenated form of haemoglobin.

33) What is the name of the collateral circulation at the base of the brain?

Circle of Willis.

34) What is deontology?

A moral theory that indicates that:

An individual has duties.

There are rules of behavior.

This theory and behavior are universalizable.

Session 28

1) Which aneurysms are associated with adult polycystic kidney disease?

2) What is the function of the prefrontal cortex in movement?

3) Which disease is characterized by cog-wheel rigidity?

4) Name four causes of haematuria.

5) Name four different upper respiratory tract infections by site of occurrence.

6) Name four signs of cerebellar ataxia.

7) Name three substances reabsorbed by tubules of the nephron.

8) Which spinal cord levels mediate the Achilles tendon reflex?

9) What is the commonest primary renal malignancy?

10) Where in the brain would you find the vermis?

11) Which system has a positive inotropic effect on the heart – the sympathetic nervous system or the parasympathetic nervous system?

12) Which spinal cord levels mediate the biceps reflex?

13) Is spastic paralysis a sign of an upper or lower motor neurone lesion?

14) Name the functions of the cerebellum that relate to movement.

15) Which spinal cord levels mediate the triceps reflex?

16) What is the initial site of action of cardiac glycosides (e.g. digoxin)?

17) Name three possible components of renal stones.

18) Voluntary movements are executed by which part of the brain?

19) What are the four subcortical nuclei in the basal ganglia?

20) What are the major neurological signs of lower motorneurone lesions?

21) Broadly speaking, there are four classes of drugs used for treating arrhythmias. What are they?

22) Is prostatic adenocarcinoma more common in the periphery or the centre of the prostate?

23) What is ataxia?

24) After contraction of a myocyte where does the calcium go?

25) Which spinal cord levels mediate the quadriceps reflex?

26) Name four drug classes used in the management of chronic heart failure.

27) Name four acquired medical conditions that might make it difficult or impossible to swallow.

28) What is the commonest primary thyroid malignancy?

Session 28 Answers

1) Which aneurysms are associated with adult polycystic kidney disease?

Berry aneurysms in the Circle of Willis.

2) What is the function of the prefrontal cortex in movement?

It generates the intention to move.

It selects the movement plan.

It converts intention into movement.

3) Which disease is characterized by cog-wheel rigidity?

Parkinson's disease.

4) Name four causes of haematuria.

Any four of:

Infection, malignancy, renal stone, trauma, glomerulonephritis and benign prostatic hypertrophy.

5) Name four different upper respiratory tract infections by site of occurrence.

Any four of:

Sinusitis, common cold, otitis media, pharyngitis, epiglottitis, laryngotracheitis and tracheitis.

6) Name four signs of cerebellar ataxia.

Any four of:

Intention tremor, dysdiadochokinesia, slurred speech, nystagmus, general ataxia and dysmetria.

7) Name three substances reabsorbed by tubules of the nephron.

Any three of:

Water, bicarbonate, glucose, urea and sodium.

8) Which spinal cord levels mediate the Achilles tendon reflex?

S1/2.

9) What is the commonest primary renal malignancy?

Renal cell carcinoma also termed clear cell renal cell carcinoma. It is actually an adenocarcinoma.

10) Where in the brain would you find the vermis?

Between the two hemispheres of the cerebellum.

11) Which system has a positive inotropic effect on the heart – the sympathetic nervous system or the parasympathetic nervous system?

Sympathetic.

12) Which spinal cord levels mediate the biceps reflex?

C5/6.

13) Is spastic paralysis a sign of an upper or lower motor neurone lesion?

Upper.

14) Name the functions of the cerebellum that relate to movement.

Maintenance of balance.

Muscle tone and posture.

Muscle co-ordination; trajectory, speed and force.

Eye movements (vestibular-ocular reflex).

Planning movements and evaluating sensory information.

Learning of motor skills.

15) Which spinal cord levels mediate the triceps reflex?

C6/7.

16) What is the initial site of action of cardiac glycosides (e.g. digoxin)?

Inhibits the Na^+/K^+-ATPase by competing for K^+ binding.

17) Name three possible components of renal stones.

Any three of:

Calcium

Magnesium

Cystine

Urate

Oxalate

Phosphate

18) Voluntary movements are executed by which part of the brain?

Primary motor cortex.

19) What are the four subcortical nuclei in the basal ganglia?

Subthalamic nucleus.

Globus pallidus.

Substantia nigra.

Striatum (caudate nucleus, putamen and nucleus accumbens).

20) What are the major neurological signs of lower motorneurone lesions?

Flaccid paralysis, absence or diminishment of spinal reflexes, muscular atrophy and muscular fasciculation.

21) Broadly speaking, there are four classes of drugs used for treating arrhythmias. What are they?

Class I – Membrane stabilizers.

Class II – β blockers.

Class III – Prolongers of the action potential.

Class IV – Calcium channel blockers.

22) Is prostatic adenocarcinoma more common in the periphery or the centre of the prostate?

Because **70% are peripheral** the per rectum examination can often identify malignancies by palpation.

23) What is ataxia?

Uncoordinated voluntary movements.

24) After contraction of a myocyte where does the calcium go?

The calcium ions are either taken up by the sarcoplasmic reticulum or are transported into the extracellular medium.

25) Which spinal cord levels mediate the quadriceps reflex?

L3/4.

26) Name four drug classes used in the management of chronic heart failure.

Any four of:

Diuretics, ACE inhibitors, Angiotensin II type 1 receptor blockers, Cardiac glycosides and β blockers.

27) Name four acquired medical conditions that might make it difficult or impossible to swallow.

Stroke, dementia, malignancy, multiple sclerosis, neurodegenerative diseases e.g. motor neurone disease.

28) What is the commonest primary thyroid malignancy?

Papillary carcinoma of the thyroid.

Session 29

1) Define nephrotic syndrome.

2) What is the embryological origin of the thyroid?

3) Name three possible treatments for hyperthyroidism.

4) What are the key clinical features of polycystic ovary disease?

5) What is a first order sensory neurone?

6) Name three microvascular complications of diabetes mellitus.

7) A patient has recently developed an increased heart rate, irritability, heat intolerance and has lost significant weight. What is the most likely diagnosis?

8) What is the clinical synonym for myxoedema?

9) State three causes of nephrotic syndrome.

10) What class of disease is Hashimoto's thyroiditis?

11) List the classical and recognized health effects of unemployment.

12) According to the WHO analgesic ladder, in general terms, what is the most effective combination of pain relieving drugs?

13) Which disease is characterized by Kimmelstiel-Wilson nodules?

14) If a patient's TSH is unusually low and free T4 is unusually high, what diagnosis does this indicate?

15) What effect will taking phenytoin have on a patient's free T4 levels?

16) Under what circumstances can a directed donation of an organ occur?

17) What is neuropathic pain?

18) What core investigations are required to define and classify glomerulonephritis on a biopsy?

19) What is the half-life of thyroxine?

20) What is the mass number for the radioactive iodine used to treat hyperthyroidism?

21) What is the current birth rate in the UK (2011)?

22) What is the function of "A delta" neurones?

23) Why are diabetics more prone to urinary tract infections?

24) Name three common causes of thyrotoxicosis.

25) List three disadvantages of postmenopausal hormonal therapy.

26) What is the normal range of total plasma calcium concentration in mmol/l?

27) In what form is T4 predominantly carried in the blood?

28) Describe what happens to the concentrations of H^+/ K^+/Na^+/ Ca^{2+}/ and PO_4^{2-} in the plasma during renal failure.

29) What is the classical presentation of a renal stone?

Session 29 Answers

1) Define nephrotic syndrome.

A glomerular lesion that results in a proteinuria of three or more grams of protein in a 24 hour urine collection. The clinical features include hypoalbuminaemia, oedema and hyperlipidaemia.

2) What is the embryological origin of the thyroid?

First branchial pouch – endodermal in origin.

3) Name three possible treatments for hyperthyroidism.

Carbimazole.

Propylthiouracil.

β blockers

Radioiodine.

Thyroidectomy.

4) What are the key clinical features of polycystic ovary disease?

Anovulation/oligoovulation.

Hyperandrogenism.

Insulin resistance/glucose intolerance.

Obesity.

Endometrial proliferation.

Infertility.

5) What is a first order sensory neurone?

The primary afferent. It responds to the stimulus, transduces the stimulus and transmits encoded information to the CNS.

6) Name three microvascular complications of diabetes mellitus.

Blindness – diabetic retinopathy.

Autonomic neuropathy – e.g. hypotension.

Peripheral neuropathy.

Foot ulcers and infection.

Diabetic nephropathy.

7) A patient has recently developed an increased heart rate, irritability, heat intolerance and has lost significant weight. What is the most likely diagnosis?

Hyperthyroidism.

8) What is the clinical synonym for myxoedema?

Hypothyroidism.

9) State three causes of nephrotic syndrome.

Any three of:

Minimal change nephropathy.

Diabetes mellitus.

Membranous glomerulonephritis.

Membranoproliferative glomerulonephritis.

Mesangioproliferative glomerulonephritis.

Focal segmental glomerulosclerosis.

Amyloidosis.

Sickle cell anaemia. IgA nephropathy.

Malaria. Systemic lupus erythematosus.

10) What class of disease is Hashimoto's thyroiditis?

Autoimmune thyroiditis.

11) List the classical and recognized health effects of unemployment.

Increased tobacco consumption.

Increased alcohol consumption.

Increased sexual health risk taking.

Increased use of GP services.

Increased use of medication.

Increased admission to psychiatric hospitals.

Increased incidence of depression.

Increased incidence of anxiety.

Increased rates of obesity.

Decreased physical activity.

Increased cardiovascular morbidity and mortality.

Decreased self-esteem.

Increased debt.

Greater difficulty in cessation of smoking.

12) According to the WHO analgesic ladder, in general terms, what is the strongest combination of pain relieving drugs?

Strong opioids and non-opioid plus an adjuvant.

13) Which disease is characterised by Kimmelstiel-Wilson nodules?

Diabetes mellitus; Kimmelstiel-Wilson nodules are deposited in the glomerulus.

14) If a patient's TSH is unusually low and free T4 is unusually high, what diagnosis does this indicate? This most likely to be primary hyperthyroidism – with T4 causing negative feedback on TSH.

15) What effect will taking phenytoin have on a patient's free T4 levels?

The free T4 levels will decrease because the induction of liver cytochrome P450 enzymes will lead to further glucuronidation and clearance of T4.

16) Under what circumstances can a directed donation of an organ occur?

A living donor can make a directed donation.

17) What is neuropathic pain?

Pain that arises as a result of nerve cell injury and so comes from within the peripheral and central nervous system.

18) What core investigations are required to define and classify glomerulonephritis on a biopsy?

Light microscopy, immunofluorescence and electron microscopy.

19) What is the half-life of thyroxine?

5-7 days.

20) What is the mass number for the radioactive iodine used to treat hyperthyroidism?

131.

21) What is the current birth rate in the UK (2011)?

1.4 – 1.8 children born per woman across her child bearing years.

22) What is the function of "A delta" neurones?

These myelinated nerves allow fast transmission of pain and temperature sensations.

23) Why are diabetics more prone to urinary tract infections?

Their glucosuria encourages bacterial growth.

Their defective neutrophil function causes an effective relative immunosuppression.

Their increased occurrence of impaired bladder evacuation predisposes to infection (autonomic neuropathy).

Their uroepithelial cells show increased adherence of bacteria.

24) Name three common causes of thyrotoxicosis.

Any three of:

Graves' disease.

Toxic multinodular goitre.

Toxic solitary adenoma.

Pituitary tumour.

Overtreatment by doctors (iatrogenic).

25) List three disadvantages of postmenopausal hormonal therapy.

Increased risk of breast carcinoma.

Increased risk of heart disease.

Increased of risk of thromboses.

Increased risk of myocardial infarction.

Increased risk of hypertension.

Increased risk of endometrial carcinoma.

Increased risk of gallstones and cholelithiasis.

26) What is the normal range of total plasma calcium concentration in mmol/l?

2.12-2.6 mmol/l.

27) In what form is T4 predominantly carried in the blood?

The majority is bound to protein; 80% is bound to thyroxine-binding globulin.

28) Describe what happens to the concentrations of H^+/ K^+/Na^+/ Ca^{2+}/ and PO_4^{2-} in the plasma during renal failure?

$[H^+]$ increases.

$[K^+]$ increases.

$[Na^+]$ decreases.

$[Ca^{2+}]$ decreases.

$[PO_4^{2-}]$ increases.

29) What is the classical presentation of a renal stone?

Spasmodic pain (*"colicky pain"*) that can pass from *loin to groin* is the classic symptom.

Haematuria, irritative urological symptoms and vomiting can also occur.

Session 30

1) Loop diuretics inhibit which transporter in the ascending loop of Henlé?

2) Where in the eye will you find the vitreous humour?

3) When are you likely to suffer from hyposmia?

4) Where in the nephron is most of the Ca^{2+} reabsorbed?

5) What class of diuretic is mannitol and when is it used?

6) According to NICE which drugs are commonly used to treat osteoporosis?

7) What type of photoreceptors exist in the retina and what are their functions?

8) What is hypogeusia?

9) What is the other name for 1,25 $(OH)_2$ D_3?

10) What effect does chronic kidney disease have on parathormone/ parathyroid hormone secretion?

11) What class of diuretic is preferentially used to manage SIADH?

12) What are the two major side effects of K^+ sparing diuretics?

13) What is astigmatism?

14) Which nerve provides the taste innervation for the posterior third of the tongue?

15) Name two symptoms of hypocalcaemia.

16) Children can suffer from rickets. What is the equivalent adult disease?

17) What is the maximum stone size that can pass unassisted through the ureters?

18) Which transporter do thiazides inhibit to cause their primary action?

19) What is the fovea?

20) Describe phantosmia.

21) Which cells make parathyroid hormone?

22) What effect can $2°$ hyperparathyroidism have on bone?

23) What are the classical and worrying clinical features of fluid overload?

24) What is the key protein target of amiloride when it is working as a K^+ sparing diuretic?

25) What is the medical term for far sightedness?

26) Which cranial nerve supplies the taste innervation for the anterior two thirds of the tongue?

27) What is the net primary effect of parathyroid hormone?

28) Sufferers from which type of disease are notorious for suffering from a β_2-microglobulin amyloidosis?

29) How does spironolactone work as a potassium sparing diuretic?

30) What is the medical term for short sightedness?

31) What are the primary modalities that form taste?

Session 30 Answers

1) Loop diuretics inhibit which transporter in the ascending loop of Henlé?

The $Na^+Cl^-K^+$ transporter.

2) Where in the eye will you find the vitreous humour?

Behind the lens and directly in front of the retina.

3) When are you likely to suffer from hyposmia?

If you have a cold, hayfever (seasonal allergic rhinitis) or sinusitis.

4) Where in the nephron is most of the Ca^{2+} reabsorbed?

60-70% at the proximal tubule.

5) What class of diuretic is mannitol and when is it used?

It is an osmotic diuretic used to decrease intracranial or intraocular pressure.

6) According to NICE which drugs are commonly used to treat osteoporosis?

Bisphosphonates, Selective oestrogen receptor modulators (SERMs), Parathyroid hormone, Strontium ranelate, Vitamin D and Ca^{2+} ion supplementation.

7) What type of photoreceptors exist in the retina and what are their functions?

Rods are responsible for black and white vision and are effective in dim lighting.

Cones – responsible for colour vision and are capable of high acuity.

8) What is hypogeusia?

Decreased sensation of taste.

9) What is the other name for $1,25\ (OH)_2\ D_3$?

Calcitriol (active vitamin D).

10) What effect does chronic kidney disease have on parathormone/ parathyroid hormone secretion?

The kidney disease decreases the level of active vitamin D that then decreases Ca^{2+} absorption and consequently lowers the Ca^{2+} concentration in the plasma.

The parathyroids increase parathyroid hormone secretion in order to return the Ca^{2+} levels to optimum physiological concentrations.

11) What class of diuretic is preferentially used to manage SIADH?

ADH antagonists.

12) What are the two major side effects of K^+ sparing diuretics?

Gynaecomastia – this anti-androgen effect is believed to be due to mineralocorticoid receptor antagonism.

Hyperkalaemia with or without acidosis.

13) What is astigmatism?

Astigmatism occurs as a result of asymmetrical curvature of the lens of the eye resulting in uneven refraction of light.

14) Which nerve provides the taste innervation for the posterior third of the tongue?

Cranial nerve IX; Glossopharyngeal nerve.

15) Name two symptoms of hypocalcaemia.

Any three of:

Seizures, muscle tetany and neuromuscular irritability. Paraesthesia around the mouth, in the mouth and in the hands and feet.

Laryngospasm and cardiac arrhythmias. Depression, irritability, personality change and fatigue.

16) Children can suffer from rickets. What is the equivalent adult disease?

Osteomalacia.

17) What is the maximum stone size that can pass unassisted through the ureters?

About 5mm in maximum extent.

18) Which transporter do thiazides inhibit to cause their primary action?

Na^+Cl^- cotransporter in the kidney.

19) What is the fovea?

It is in the centre of the retina; the area of greatest visual acuity.

20) Describe phantosmia.

Olfactory hallucinations. Smelling an odour that is not present.

21) Which cells make parathyroid hormone?

Chief cells of the parathyroid glands.

22) What effect can $2°$ hyperparathyroidism have on bone?

It causes increased bone resorption. It causes formed bone to be disorganized and weak. It causes the deposition of fibrous tissue.

23) What are the classical and worrying clinical features of fluid overload?

Pulmonary oedema/peripheral oedema.

Congestive cardiac failure.

Hypertension.

24) What is the key protein target of amiloride when it is acting as a K^+ sparing diuretic?

Amiloride blocks the luminal epithelial sodium channel ENaC in the nephron.

25) What is the medical term for far sightedness?

Hypermetropia.

26) Which cranial nerve supplies the taste innervation for the anterior two thirds of the tongue?

Cranial nerve VII which is the *Facial* nerve.

27) What is the net primary effect of parathyroid hormone?

Raises the serum Ca^{2+} concentration.

28) Sufferers from which type of disease are notorious for suffering from a β_2-microglobulin amyloidosis?

Any individual on **long-term dialysis**, usually for **chronic renal failure**. β_2-microglobulin is not well filtered.

29) How does spironolactone work as a potassium sparing diuretic?

It binds to the renal aldosterone receptor. Which then:

(a) Inhibits Na^+ reabsorption at the luminal sodium channel, leading to a diuresis.

(b) Inhibits K^+ excretion (combined effects at the Na^+/K^+ pump and the K^+ channel).

30) What is the medical term for short sightedness?

Myopia.

31) What are the primary modalities that form taste?

Sweet, salt, sour, bitter and savoury.

Session 31

1) What is the threshold age below which menopause is considered premature?

2) If emphysema is present and distributed equally throughout the bronchioles and alveoli, what morphological type of emphysema is it?

3) What is the target HbA1c level in a well-controlled diabetic?

4) The sympathetic innervation of the bladder originates from which level?

5) Surgery to the parotid gland may traumatize which cranial nerve?

6) Where would you expect to find the Glut-4 transporter?

7) The parasympathetic innervation of the bladder originates from which levels?

8) Name the biggest risk factor for chronic obstructive pulmonary disease.

9) Which is more likely to predispose to atherosclerosis:
   a) Raised HDL and low LDL *or*
   b) Low HDL and raised LDL

10) What is the mechanism of action of a sulphonylurea?

11) What is the clinical definition of chronic bronchitis?

12) Name the first line drug used to treat urinary tract infections.

13) What is metformin's primary mechanism of action?

14) Name two possible causes of an atonic bladder.

15) What is Prempak C?

16) What are the two components of chronic obstructive pulmonary disease?

17) Name the first choice cholesterol lowering drug in a man over 40 with a family history of IHD and a personal history of diabetes mellitus.

18) What is the alphabet strategy for managing diabetes mellitus?

19) Does stimulation of β adrenergic receptors in the body of the bladder decrease or increase detrusor muscle tone?

20) Name the three clinical classes of insulin.

21) Name the branches of the facial nerve.

22) Where would you expect to find the Glut-3 transporter?

23) When are you permitted to register with the GMC, after F1 or after F2?

24) Name the biggest host risk factor for chronic obstructive pulmonary disease.

25) How does the average life expectancy of a diabetic compare to that of a non-diabetic?

26) What is the trigone and how would you identify it anatomically?

27) Define menopause.

Session 31 Answers

1) What is the threshold age below which menopause is considered premature?

45 years of age.

2) If emphysema is present and distributed equally throughout the bronchioles and alveoli, what morphological type of emphysema is it?

Panacinar.

3) What is the target HbA1c level in a well-controlled diabetic?

≤ 7% of HbA1c

4) The sympathetic innervation of the bladder originates from which level?

T11-L2.

5) Surgery to the parotid gland may traumatize which cranial nerve?

Facial nerve; cranial nerve VII.

6) Where would you expect to find the Glut-4 transporter?

Skeletal muscle cells and adipose tissue (sites of insulin stimulated glucose uptake).

7) The parasympathetic innervation of the bladder originates from which levels?

S2-S4.

8) Name the biggest risk factor for chronic obstructive pulmonary disease.

Smoking.

9) Which is more likely to predispose to atherosclerosis:
   a) Raised HDL and low LDL *or*
   b) Low HDL and raised LDL

b) Low HDL and raised LDL.

10) What is the mechanism of action of a sulphonylurea?

It increases insulin secretion by binding to K^+ channels at the cell membrane of beta cells in the Islets of Langerhans. The depolarizing effect on the membrane potential opens voltage gated calcium channels. The influx of Ca^{2+} is needed for and facilitates the insulin exocytosis. The insulin then acts as an endocrine hormone.

11) What is the clinical definition of chronic bronchitis?

A chronic productive cough for at least three months in two successive years (assuming other pathologies have been excluded).

12) Name the first line drug used to treat urinary tract infections.

Trimethoprim.

13) What is metformin's primary mechanism of action?

It inhibits *hepatic gluconeogenesis*, thus decreasing the secretion of glucose into the circulation. (It also causes an increase in insulin sensitivity, increases peripheral glucose uptake and increases fatty acid oxidation).

14) Name two possible causes of an atonic bladder.

There are two major groups of causes:

a) **(Peripheral) nerve damage** – diabetes mellitus, multiple sclerosis and trauma e.g. pelvic surgery/spinal cord damage, are common causes of nerve damage.

b) **Obstructive uropathy** – benign prostatic hypertrophy and neoplasms are common causes of obstructing neuropathy.

15) What is Prempak C?

It is a post menopausal hormone replacement therapy that contains both an oestrogen and a progestogen.

16) What are the two components of chronic obstructive pulmonary disease?

Emphysema and chronic bronchitis.

17) Name the first choice cholesterol lowering drug in a man over 40 with a family history of IHD and a personal history of diabetes mellitus.

Currently simvastatin is the first choice statin.

18) What is the alphabet strategy for managing diabetes mellitus?

Advice =>	Smoking/Diet/Exercise.
Blood pressure =>	≤140/80
Cholesterol =>	Total Cholesterol ≤4 LDL:HDL ≤2
Diabetes control =>	HbA1c ≤7%
Eye examination =>	Annual examination
Feet examination =>	Annual examination
Guardian drugs =>	Aspirin/ACEI/Statins *et al.*

19) Does stimulation of β adrenergic receptors in the body of the bladder decrease or increase detrusor muscle tone?

Decrease.

20) Name the three clinical classes of insulin.

Short acting, intermediate acting and long acting insulins.

21) Name the branches of the facial nerve.

Temporal, zygomatic, buccal, mandibular and cervical.

22) Where would you expect to find the Glut-3 transporter?

The brain.

23) When are you permitted to register with the GMC, after F1 or after F2?

After F1, the end of your first year practising as a doctor.

24) Name the biggest host risk factor for chronic obstructive pulmonary disease.

α-1-antitrypsin deficiency.

25) How does the average life expectancy of a diabetic compare to that of a non-diabetic?

Currently approximately 5-10 years less for a diabetic.

26) What is the trigone and how would you identify it anatomically?

It is the part of the bladder wall that is between the ureteric openings and the urethral exit. The mucosa is smooth.

27) Define menopause.

Cessation of menstruation for at least one year; it is accompanied by the end of ovulation and decreased levels of oestrogens.

Session 32

1) What is the commonest cause of death in this country (UK)?

2) Name four paraneoplastic signs.

3) What type of malignancy of the testis is a man between the ages of 30-50 most likely to have?

4) What is the classical test for the level of consciousness that doctors use to assess patients?

5) A Berry aneurysm ruptures. What is the term for the subsequent bleed?

6) What type of granuloma is characteristic of TB?

7) What is the correlation between domestic violence and child abuse?

8) What does palliate mean?

9) Name the three types of management of ADHD (Attention Deficit Hyperactivity Disorder) and list two examples of each type.

10) What type of malignancy of the testis is a man aged 20-30 most likely to have?

11) Which artery typically causes an extradural haemorrhage?

12) What are the four drugs usually used in TB chemotherapy?

13) A normal state of consciousness depends on the balance of which three CNS factors?

14) What % of women have been physically assaulted by their partner at some point?

15) Name four risk factors for bladder cancer.

16) What is the name for renal cell carcinoma metastases to the lung that cause round discrete lesions on a chest X-ray image?

17) What type of testicular malignancy is a man of 50+ most likely to have?

18) What is the normal capacity of a human bladder?

19) What does the acronym NICE stand for?

20) What are the risk factors for sleep apnoea?

21) Name five bladder tumours.

22) What type of malignancy is the commonest primary prostate malignancy?

23) What is the commonest type of penile malignancy?

24) What is the commonest underlying cause of intraparenchymal haemorrhage of the brain?

25) Name three different tests for TB (which may be any clinical, microbiological or pathological tests).

26) Name five types of incontinence.

27) A patient has an MI and is treated with streptokinase. Three days later another MI occurs. As a result, what are the current treatment alternatives?

Session 32 Answers

1) What is the commonest cause of death in this country (UK)?

"Cardiovascular disease" usually *ischaemic heart disease*.

2) Name four paraneoplastic signs.

Any four of:

Anaemia, erythrocytosis, cachexia, hepatic dysfunction, hormonal abnormalities and hypercalcaemia/hypercalcaemic signs.

3) What type of malignancy of the testis is a man between the ages of 30-50 most likely to have?

Seminoma.

4) What is the classical test for the level of consciousness that doctors use to assess patients?

Glasgow Coma Scale.

5) A Berry aneurysm ruptures. What is the term for the subsequent bleed?

A subarachnoid haemorrhage.

6) What type of granuloma is characteristic of TB?

Caseating granuloma (or Caseous granuloma).

7) What is the correlation between domestic violence and child abuse?

There is a positive correlation of between 30-60%.

8) What does palliate mean?

To alleviate a disease without curing it.

9) Name the three types of management of ADHD (Attention Deficit Hyperactivity Disorder) and list two examples of each type.

Pharmacological – Ritalin, Dexedrine or Strattera (Atormoxetine).

Therapy – Psychotherapy, Behavioural Therapy, Parenting Skills and Social Skills Training.

Dietary – High protein diet and carbohydrate replacement.

10) What type of malignancy of the testis is a man aged 20-30 most likely to have?

Teratoma.

11) Which artery typically causes an extradural haemorrhage?

Middle meningeal artery. (An extradural haematoma is sometimes termed an *epidural* haematoma).

12) What are the four drugs usually used in TB chemotherapy?

Rifampicin, isoniazid, pyrazinamide and ethambutol.

13) A normal state of consciousness depends on the balance of which three CNS factors?

Alertness – reticular formation.

Attention – limbic system and frontoparietal association areas.

Awareness – cerebral cortex.

14) What % of women have been physically assaulted by their partner at some point?

Approximately 20%.

15) Name four risk factors for bladder cancer.

Increasing age.

Being male.

Smoking.

Aniline dye workers.

Schistosomiasis infection.

Drugs – phenacetin/cyclophosphamide.

16) What is the name for renal cell carcinoma metastases to the lung that cause round discrete lesions on a chest X-ray image?

Cannonball metastases.

17) What type of testicular malignancy is a man of 50+ most likely to have?

Lymphoma.

18) What is the normal capacity of a human bladder?

300 - 600mls.

19) What does the acronym NICE stand for?

National Institute of Clinical Excellence.

20) What are the risk factors for sleep apnoea?

Increased weight.

Neck circumference greater than 17cm.

Being male.

High blood pressure.

Enlarged tonsils or adenoids.

Obstructing polyp.

Increasing age.

Family history.

Use of alcohol or sedatives.

Smoking.

21) Name five bladder tumours.

Primary tumours include transitional cell carcinoma, transitional cell carcinoma in situ, squamous cell carcinoma, adenocarcinoma and sarcoma. Secondary tumours include adenocarcinoma, lymphoma and small cell carcinoma.

22) What type of malignancy is the commonest primary prostate malignancy?

Adenocarcinoma.

23) What is the commonest type of penile malignancy?

Squamous cell carcinoma.

24) What is the commonest underlying cause of intraparenchymal haemorrhage of the brain?

Hypertension.

25) Name three different tests for TB (which may be any clinical, microbiological or pathological tests).

Mantoux test.

Ziehl-Neelsen stain.

Microbiological culture.

T-spot Elispot test.

26) Name five types of incontinence.

Stress.

Urge.

Overflow.

Functional.

Continuous.

27) A patient has an MI and is treated with streptokinase. Three days later another MI occurs. As a result, what are the current treatment alternatives?

tPA or coronary angioplasty.

Session 33

1) Name four primary lung malignancies.

2) Which is responsible for bladder contraction, the sympathetic or parasympathetic nervous system?

3) Haemoptysis is a sign of which type of tumour?

5) What is visual agnosia?

6) Name two radiological investigations that are commonly used to determine the presence of widespread metastatic dissemination.

7) What is the term for loss of memory of events occurring *prior to a CNS trauma*?

8) What is another phrase that means the same as "grand mal seizure"?

9) Which lung malignancy is notoriously susceptible to chemotherapy?

10) Which disorder can tegretol, epilim and valium be used to manage?

11) What is sensitivity?

12) What is extracorporeal shock wave lithotripsy usually used to treat?

13) Name a chemotherapeutic alkylating agent.

14) Which disease is associated with granulomas, Crohn's disease or ulcerative colitis?

15) Which organs account for insensible fluid loss?

16) Name four natural processes that raise intraabdominal pressure.

17) What is specificity?

18) What type of collagen would you expect to find in tendons or ligaments?

19) What is achalasia?

20) Does fibre increase or decrease gastrointestinal transit time?

21) Is lung adenocarcinoma more common in the perihilar region or in the periphery?

22) Poor glycaemic control is now known to be a major factor in determining whether a diabetic will suffer vascular or neuropathic complications. Apart from glucose, **which sugar** accumulates and has an aetiological role in complications, under conditions of poor glycaemic control?

Session 33 Answers

1) Name four primary lung malignancies.

Any four of:

Squamous cell carcinoma, adenosquamous carcinoma, *adenocarcinoma*, bronchioloalveolar carcinoma, *small cell carcinoma* (oat cell carcinoma), large cell carcinoma, carcinoid tumour, pleuropulmonary blastoma, sarcomatoid carcinoma and salivary gland tumours.

2) Which is responsible for bladder contraction, the sympathetic or parasympathetic nervous system?

Parasympathetic.

3) Haemoptysis is a sign of which type of tumour?

Lung neoplasm (or lung malignancy).

5) What is visual agnosia?

Inability to recognize familiar objects.

6) Name two radiological investigations that are commonly used to determine the presence of widespread metastatic dissemination.

Any two of:

Isotopic bone scan, MRI scan or PET scan.

7) What is the term for loss of memory of events occurring prior to a CNS trauma?

Retrograde amnesia.

8) What is another phrase that means the same as "grand mal seizure"?

Generalized clonic-tonic seizure.

9) Which lung malignancy is notoriously susceptible to chemotherapy?

Small cell carcinoma.

10) Which disorder can tegretol, epilim and valium be used to manage?

Epilepsy. (Tegretol is carbamezapine, epilim is sodium valproate and valium is diazepam).

11) What is sensitivity?

Sensitivity is the proportion of people who have the disease that the test correctly detects.

Sensitivity = True positives/(True positives + False negatives)

12) What is extracorporeal shock wave lithotripsy usually used to treat?

Renal stones.

13) Name a chemotherapeutic alkylating agent.

Common examples are:

Cyclophosphamide, Busulphan, Melphalan, Lomustine, Carmustine and Thiotepa.

14) Which disease is associated with granulomas, Crohn's disease or ulcerative colitis?

Crohn's disease.

15) Which organs account for insensible fluid loss?

Skin and lungs.

16) Name four natural processes that raise intraabdominal pressure.

Any four of:

Straining (defaecation), coughing, sneezing, micturition, parturition and laughing.

17) What is specificity?

Specificity is the proportion of people who do not have the disease that the test correctly identifies as not having the disease.

Specificity = True Negatives/(True negatives + False positives)

18) What type of collagen would you expect to find in tendons or ligaments?

Type 1 collagen.

19) What is achalasia?

Achalasia is disorder of motility of oesophageal muscle that leads to dysphagia.

20) Does fibre increase or decrease gastrointestinal transit time?

Decrease; fibre enhances movement of luminal contents through the gastrointestinal tract.

21) Is lung adenocarcinoma more common in the perihilar region or in the periphery?

Lung periphery (there is more lung parenchyma in the periphery than in the hilum).

22) Poor glycaemic control is now known to be a major factor in determining whether a diabetic will suffer vascular or neuropathic complications. Apart from glucose, which sugar accumulates and has an aetiological role in complications, under conditions of poor glycaemic control?

Sorbitol production is increased when blood glucose rises as a result of the polyol pathway. Sorbitol is associated with increased glycosylation of proteins and increased inflammation.

Session 34

1) What type of epithelium lines the bronchi and trachea?

2) Name three different pneumoconioses.

3) Name three complications of reflux oesophagitis.

4) Name three synonymous phrases that have been used for *autism*.

5) Describe retrograde amnesia.

6) What are the major complications of chronic kidney disease that can directly involve the foetus?

7) Name the acronym you use to remember the key components of a patient's past medical history.

8) What class of disease is Bird fancier's lung?

9) Which lung disease is classically characterized by fibroblastic foci?

10) Which is more likely to contain bacteria, a parapneumonic effusion or an empyema?

11) Which contains more protein, a transudate or an exudate?

12) Name the gastrointestinal disease that is classically associated with erythema nodosum.

13) Name four classes of purgatives.

14) How does creatinine clearance change with age?

15) Explain the term constipation.

16) Name two proton pump inhibitor drugs.

17) What is rivastigmine?

18) What kind of dysphasia is caused by damage to Broca's area?

19) Which part of the cortex is responsible for the mood/personality?

20) Which dysphasia is associated with fluent excessive speech, lack of insight, jargon use and poor comprehension?

21) Name two synonyms for *cryptogenic fibrosing alveolitis.*

22) What is azotorrhoea?

23) Explain what is meant by a type III hypersensitivity reaction.

Session 34 Answers

1) What type of epithelium lines the bronchi and trachea?

Respiratory epithelium which is **ciliated pseudostratified columnar epithelium.**

2) Name three different pneumoconioses.

Common answers include:

Coal worker's pneumoconiosis.

Caplan's syndrome.

Silicosis.

Asbestosis.

Berylliosis.

Aluminium/Bauxite fibrosis.

Siderosis (iron).

Baritosis (barium).

Stannosis (tin).

3) Name three complications of reflux oesophagitis.

Complications of reflux oesophagitis include:

Barrett's oesophagus.

Oesophageal carcinoma.

Oesophageal stricture.

Gastrointestinal bleeding.

Iron deficient anaemia.

Aspiration pneumonia.

4) Name three synonymous phrases that have been used for *autism*.

Asperger syndrome.

High functioning autism.

Autism spectrum disorder.

Autism spectrum condition.

Autism spectrum.

5) Describe retrograde amnesia.

Impairment of memory for events that antedate the trauma or injury.

6) What are the major complications of chronic kidney disease that can directly involve the foetus?

Prematurity.

Small for age.

Foetal death.

7) Name the acronym you use to remember the key components of a patient's past medical history. **There are several possible acronyms, however a popular one is MIJ THREADS.**

MI	Myocardial infarction
J	Jaundice
T	Tuberculosis
H	Hypertension
R	Rheumatic fever
E	Epilepsy
A	Asthma
D	Diabetes mellitus
S	Stroke

Hence you should remember to ask if the patient has had a MI, Jaundice, TB etc.....

8) What class of disease is Bird fancier's lung?

The best answer is an **extrinsic allergic alveolitis**. Other possibilities include granulomatous disease, hypersensitivity reaction or immunological disease.

9) Which lung disease is classically characterized by fibroblastic foci?

Idiopathic pulmonary fibrosis/usual interstitial pneumonia/cryptogenic fibrosing alveolitis.

10) Which is more likely to contain bacteria, a parapneumonic effusion or an empyema?

An empyema.

11) Which contains more protein, a transudate or an exudate?

An exudate.

12) Name the gastrointestinal disease that is classically associated with erythema nodosum.

Ulcerative colitis has a stronger association with erythema nodosum than Crohn's disease.

13) Name four classes of purgatives.

Bulk laxatives, osmotic laxatives, faecal softeners and stimulant purgatives.

14) How does creatinine clearance change with age?

Creatinine clearance decreases with age.

15) Explain the term constipation.

It refers to infrequent or difficult evacuation of faeces that usually involves straining. There are less than or equal to three bowel movements per week.

16) Name two proton pump inhibitor drugs.

Any two of:

Omeprazole, pantoprazole, lansoprazole, esomeprazole or rabeprazole.

17) What is rivastigmine?

An acetylcholinesterase inhibitor.

18) What kind of dysphasia is caused by damage to Broca's area?

Expressive dysphasia.

19) Which part of the cortex is responsible for the mood/personality?

Prefrontal cortex.

20) Which dysphasia is associated with fluent excessive speech, lack of insight, jargon use and poor comprehension?

Wernicke's dysphasia

21) Name two synonyms for *cryptogenic fibrosing alveolitis*.

Cryptogenic Fibrosing Alveolitis

= Fibrosing alveolitis

= Idiopathic pulmonary fibrosis

= Diffuse fibrosing alveolitis

= Usual interstitial pneumonia/pneumonitis

= Interstitial diffuse pulmonary fibrosis

= Alveolocapillary block

= Hamman – Rich syndrome

22) What is azotorrhoea?

Faeces that show protein malabsorption and so contain excess nitrogenous breakdown products that produce a putrid aroma.

23) Explain what is meant by a type III hypersensitivity reaction.

Type III hypersensitivity reactions are characterized by antigen-antibody complexes that induce inflammation or activate complement. The aetiology of systemic lupus erythematosus is a type III hypersensitivity reaction.

Session 35

1) Which medical emergency makes use of calcium resonium?

2) Name two strong opioid analgesics.

3) Name four adverse effects of opioids.

4) What is the commonest cause of superior vena caval obstruction?

5) Considering health economics, what is *cost benefit*?

6) If a patient has acute renal impairment what life threatening complications may result?

7) Name four different NSAIDS (non-steroidal anti-inflammatory drugs) in common use.

8) What is drug tolerance?

9) What is the difference in management between small cell and non-small cell lung cancer?

10) After bad news has been broken to a patient what information does a patient usually ask for?

11) What is the first step in the drug management of hyperkalaemia?

12) What types of clinical uses do NSAIDS have?

13) Which drug would you use to reverse an opioid overdose?

14) Define cost utility.

15) What are the most appropriate situational circumstances for breaking bad news?

16) Considering neurology what does DAI stand for?

17) Where in the body are opioids broken down, primarily?

18) Which lung malignancies are caused by smoking (or increased in prevalence by smoking)?

19) What is evidence-based medicine?

20) What is the common name for acetaminophen?

Session 35 Answers

1) Which medical emergency makes use of calcium resonium?

Hyperkalaemia (calcium resonium chelates potassium ions).

2) Name two strong opioid analgesics.

Any two of:

Methadone.

Fentanyl.

Morphine.

Diamorphine.

Oxymorphone.

Oxycodone.

Hydrocodone.

3) Name four adverse effects of opioids.

Respiratory depression.

Mood alterations.

Reduced gastric motility.

Nausea and vomiting.

Miosis.

Smooth muscle spasm.

Anaphylaxis.

4) What is the commonest cause of superior vena caval obstruction?

Malignancy.

5) Considering health economics, what is *cost benefit*?

Health cost and benefits expressed in monetary units.

6) If a patient has acute renal impairment what life threatening complications may result?

Hyperkalaemia.

Severe uraemia.

Severe metabolic acidosis.

GI haemorrhage.

Pulmonary oedema.

Sepsis.

7) Name four different NSAIDS (non-steroidal anti-inflammatory drugs) in common use.

Any four of:

Aspirin

Diclofenac

Indomethacin

Ibuprofen

Piroxicam

Mefenamic acid

Celecoxib

Naproxen

8) What is drug tolerance?

Drug tolerance occurs as habituation to the drug requires an increasing dose to produce the same original effect. Similarly, the same dose of the drug produces a diminishing effect on repeated administration.

9) What is the difference in management between small cell and non-small cell lung cancer?

The most significant difference in management is the possibility of cure of the small cell carcinoma by chemotherapy (approximately a 30% 5 year survival). In contrast the management of non-small cell carcinoma usually involves surgical excision, if possible, and accompanying radiotherapy.

10) After bad news has been broken to a patient what information does a patient usually ask for?

Information about diagnosis.

Chances of cure.

Treatment side effects.

Prognosis.

11) What is the first step in the drug management of hyperkalaemia?

The priority is to stabilize electrically excitable tissues – particularly the myocardium. The 10mls of 10% calcium gluconate is usually given intravenously as the first step in management to stabilize the myocardium.

12) What types of clinical uses do NSAIDS have?

As:

Anti-inflammatories.

Anti-pyretics.

Analgesics.

Anti-coagulants.

13) Which drug would you use to reverse an opioid overdose?

Naloxone or naltrexone.

14) Define cost utility.

It is the ratio of the cost in monetary units to an overall measure of health status. An example would be the cost per quality adjusted life year (QALY).

15) What are the most appropriate situational circumstances for breaking bad news?

Privacy.

Appropriate pace of information delivery – without distractions.

Time for questions.

Availability of supportive individuals.

16) Considering neurology what does DAI stand for?

Diffuse axonal injury.

17) Where in the body are opioids broken down, primarily?

The liver.

18) Which lung malignancies are caused by smoking (or increased in prevalence by smoking)?

All of the major primary lung cancers and mesothelioma.

(The only exceptions are bronchioloalveolar carcinoma and carcinoid tumour of the lung.)

19) What is evidence-based medicine?

"The conscientious, judicious and explicit use of current best evidence when making decisions about the care of individual patients" or
"Evidence-based medicine is the use of mathematical estimates of the risk, benefit and harm, derived from high-quality research on population samples, to inform clinical decision making in the diagnosis, investigation or management of individual patients."

20) What is the common name for acetaminophen?

Paracetamol.

Session 36

1) A biopsy from the brain of an individual with Alzheimer's disease is likely to show which histological features?

2) Which neurotransmitter levels do benzodiazepines affect?

3) Name four common risk factors for chronic kidney disease.

4) Describe the sort of individuals who are most likely to suffer from depression.

5) What neurotransmitter levels do MAOIs affect?

6) What are the advantages of X-ray imaging?

7) What does the acronym ESRF stand for?

8) When you are clinically assessing the risk of pulmonary embolism which three questions should you ask yourself?

9) What are the symptoms of depression?

10) Name five signs that are characteristic of a pulmonary embolism.

11) To what lung disorder do antithrombin III deficiency, Protein S deficiency and Factor V Leiden mutations predispose?

12) What class of drugs has tardive dyskinesia as a side effect after long-term use?

13) What is the theory of mind?

14) What neurotransmitter levels do SSRIs affect?

15) What colour is metal on an x-ray image?

16) Name two disadvantages (negative features) of CT imaging.

17) What is a reactive attachment disorder and why does it occur?

18) Which neurodegenerative disease is characterised by choreiform (dancing) movements?

19) What percentage of the population suffer from depression?

20) What percentage of patients with a deep venous thrombosis develop a pulmonary embolism?

21) To cause their antidepressant effects, which neurotransmitters do tricyclic antidepressants primarily affect?

22) What are the disadvantages of standard x-ray imaging?

23) What are the common clinical features of Alzheimer's disease?

24) What condition is clozapine most often used to manage?

25) What is the classic cause of Korsakoff's psychosis?

26) What does the phrase *affective disorder* mean?

27) What is anhedonia?

28) NSAIDS, TB and aminoglycosides can cause a similar pattern of renal impairment. What is the name of this pattern of renal impairment?

29) Which disease is cANCA positive during exacerbations and what does the acronym cANCA mean?

30) Which neurodegenerative disorder is characterized by a festinant gait and a pill-rolling tremor?

Session 36 Answers

1) A biopsy from the brain of an individual with Alzheimer's disease is likely to show which histological features?

Amyloid plaques.

Neurofibrillary tangles.

2) Which neurotransmitter levels do benzodiazepines affect?

Gamma aminobutyric acid (GABA).

3) Name four common risk factors for chronic kidney disease.

Any four of:

Increasing age.

Hypertension

Diabetes mellitus.

Ethnicity.

Male.

Smoking.

NSAID use.

4) Describe the sort of individuals who are most likely to suffer from depression.

Female.

Greater than 28 years of age.

Family history of depression.

Suffering from life stresses.

5) What neurotransmitter levels do MAOIs affect?

Noradrenaline and serotonin.

6) What are the advantages of X-ray imaging?

Fast, cheap, good bone detail and has the possibility of dynamic images.

7) What does the acronym ESRF stand for?

End stage renal failure.

8) When you are clinically assessing the risk of pulmonary embolism which three questions should you ask yourself?

a) Does the patient display the clinical syndrome of pulmonary embolism?

b) Does the patient have a major risk factor for a pulmonary embolism?

c) Is a pulmonary embolism more likely than any of the differential diagnoses in this patient?

9) What are the symptoms of depression?

Low mood, anhedonia, fatigue, sleep disturbances, psychomotor disturbances, confusion, cognitive deficits, guilt, pessimism, low self-esteem, suicidal tendencies, weight change and alterations in food intake.

10) Name five signs that are characteristic of a pulmonary embolism.

Any five of:

Acute dyspnea, haemoptysis, engorged neck veins, hypoxia, cough, wheeze, decreased cardiac output, tachycardia, hypotension, shock, cardiac arrest and collapse.

11) To what lung disorder do antithrombin III deficiency, Protein S deficiency and Factor V Leiden mutations predispose?

Pulmonary embolism.

12) What class of drugs has tardive dyskinesia as a side effect after long-term use?

Antipsychotic medications; neuroleptics.

13) What is the theory of mind?

The innate ability of one person to sense the mental state of another.

14) What neurotransmitter levels do SSRIs affect?

 Serotonin.

15) What colour is metal on an x-ray image?

White.

16) Name two disadvantages (negative features) of CT imaging.

Expensive equipment.

Very high X-ray radiation dose.

17) What is a reactive attachment disorder and why does it occur?

Reactive attachment disorder is a rare but serious condition in which infants and young children do not establish healthy bonds with parents or caregivers. Individuals suffering from reactive attachment disorder:

(a) Do not trust other people.

(b) Depend only on themselves.

(c) Keep an emotional distance from others to "stay safe."

Reactive attachment disorder is caused by pathological parenting in early life.

18) Which neurodegenerative disease is characterised by choreiform (dancing) movements?

Huntington's disease.

19) What percentage of the population suffer from depression?

Approximately 5%.

20) What percentage of patients with a deep venous thrombosis develop a pulmonary embolism?

Approximately 33%.

21) To cause their antidepressant effects, which neurotransmitters do tricyclic antidepressants primarily affect?

Noradrenaline and serotonin.

22) What are the disadvantages of standard x-ray imaging?

Only 2 dimensional.

Poor soft tissue detail.

High X-ray dose.

23) What are the common clinical features of Alzheimer's disease?

Memory loss.

Gradual behavioural changes.

Dementia.

Mean life expectancy from diagnosis is seven years.

24) What condition is clozapine most often used to manage?

Schizophrenia.

25) What is the classic cause of Korsakoff's psychosis?

Alcohol abuse.

26) What does the phrase *affective disorder* mean?

These are mood disorders and include anxiety disorders, phobias, depression and mania.

27) What is anhedonia?

Anhedonia is the inability to gain pleasure from usually enjoyable experiences/activities.

28) NSAIDS, TB and aminoglycosides can cause a similar pattern of renal impairment. What is the name of this pattern of renal impairment?

Acute interstitial nephritis (a subtype of acute renal failure).

29) Which disease is cANCA positive during exacerbations and what does the acronym cANCA mean?

Wegener's granulomatosis shows particular cANCA positivity during exacerbations, reaching a sensitivity of approximately 96%. Wegener's granulomatosis is characterized by a triad of granulomatous lesions in the upper respiratory tract, the lower respiratory tract and the kidneys. cANCA = cytoplasmic anti-neutrophil cytoplasmic antibodies.

30) Which neurodegenerative disorder is characterized by a festinant gait and a pill-rolling tremor?

Parkinson's disease.

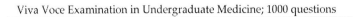

Lightning Source UK Ltd.
Milton Keynes UK
UKOW040733020113

204263UK00001B/75/P